Electromagnetics
Lecture Notes 2019

Thomas Paul Weldon

CONTENTS

PREFACE

The collection of lecture notes in this book are based on over 20 years of teaching graduate and undergraduate courses at the University of North Carolina at Charlotte. As lecture notes, this book is not intended to be a substitute for the many excellent textbooks in this field. Instead, this book is intended as a supplement to other course materials and as a workbook for students taking notes during corresponding lectures. In addition, practicing engineers may find this book useful for quick review of the topic.

Thomas Paul Weldon
Charlotte, NC
February 6, 2019

1 WAVE EQUATION AND LC TRANSMISSION LINES

The lecture notes in this chapter provide an introduction to wave equations and LC transmission lines.

Unit circle

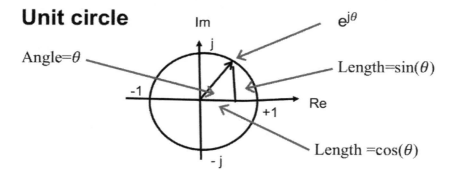

- $e^{j\theta} = \cos(\theta) + j\sin(\theta) = \alpha + j\beta$
- Can be viewed as the point in complex plane with coordinates ($\cos\theta$, $\sin\theta$)
- Can also be viewed as a vector, as shown above
- $| e^{j\theta} | = ?$

Unit Circle

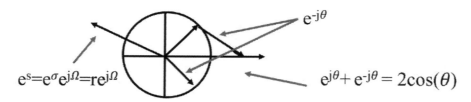

- As shown above, vector viewpoint is useful when adding
- Phasor notation can also be useful in polynomials
- Also e^s or e^{st} can be shown as above, where $s = \sigma + j\Omega$
- Consider finding square root of j

$$(\gamma \angle \theta)^2 = \gamma^2 \angle 2\theta = j = 1\angle 90°$$

or $(\gamma\, e^{j\theta})^2 = \gamma^2\, e^{j2\theta} = j = 1\, e^{j\pi/2}$ (assuming γ is positive real)

- So, there are 2 roots to 2nd order polynomial

$$1\angle 45° \text{ AND } 1\angle 225°$$

Other Useful Functions

step: $u(t) = \begin{cases} 1 & \text{if } t \geq 0 \\ 0 & \text{otherwise} \end{cases}$

pulse: $\Pi(t) = rect(t) = \begin{cases} 1 & \text{if } |t| \leq 0.5 \\ 0 & \text{otherwise} \end{cases}$

impulse: $\delta(t) = \lim_{\varepsilon \to 0}\left(\frac{\Pi(t/\varepsilon)}{\varepsilon}\right)$, and $\int_{-\infty}^{t} \delta(\alpha)d\alpha = u(t)$

triangle: $\Delta(t) = \begin{cases} 1 - 2|t| & \text{if } -0.5 \leq t \leq 0.5 \\ 0 & \text{otherwise} \end{cases}$

sinc: $sinc(t) = \frac{\sin(t)}{t}$

Wave Equations and LC Transmission Lines

(Telegrapher's Equations)

LC Transmission Lines

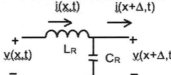

Lossless LC transmission line model

- Consider the lumped-element model of a transmission line shown above, representing an incremental length Δx

- Next, derive Telegrapher's equations governing the behavior

$$v(x,t) - \Delta x \cdot L_R \frac{\partial i(x,t)}{\partial t} = v(x+\Delta x,t) \implies \frac{v(x,t) - v(x+\Delta x,t)}{\Delta x} = L_R \frac{\partial i(x,t)}{\partial t}$$

$$i(x,t) - \Delta x \cdot C_R \frac{\partial v(x+\Delta,t)}{\partial t} = i(x+\Delta x,t) \implies \frac{i(x,t) - i(x+\Delta x,t)}{\Delta x} = C_R \frac{\partial v(x,t)}{\partial t}$$

- Where units are L_R in H/m and C_R in F/m and taking the limit $\Delta x \implies 0$ yields:

$$-\frac{\partial v(x,t)}{\partial x} = L_R \frac{\partial i(x,t)}{\partial t}$$

$$-\frac{\partial i(x,t)}{\partial x} = C_R \frac{\partial v(x,t)}{\partial t}$$

LC Transmission Lines: Telegrapher's Eqns.

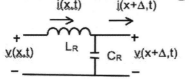

Lossless LC transmission line model

- These are Telegrapher's lossless transmission line equations:

$$\frac{\partial v(x,t)}{\partial x} = -L_R \frac{\partial i(x,t)}{\partial t} \qquad \frac{\partial i(x,t)}{\partial x} = -C_R \frac{\partial v(x,t)}{\partial t}$$

- To solve these equations, take $\partial/\partial x$ of both sides of the left equation

$$\frac{\partial^2 v(x,t)}{\partial x^2} = -L_R \frac{\partial}{\partial t}\frac{\partial}{\partial x} i(x,t)$$

- And substituting for $\partial i(x,t)/\partial t$ yields the wave equation:

$$\frac{\partial^2 v(x,t)}{\partial x^2} = L_R C_R \frac{\partial^2 v(x,t)}{\partial t^2} \longleftarrow$$

Lossless wave equation for lossless LC transmission line

See: Weldon, et al., "'Left-Handed Extensions of Maxwell's Equations for Metamaterials," IEEE SoutheastCon 2010 Proceedings, March 18-21, 2010.

Propagation in Lossless LC Transmission Lines

$$\frac{\partial^2 v(x,t)}{\partial x^2} = L_R C_R \frac{\partial^2 v(x,t)}{\partial t^2}$$

Here showing the most common engineering formats

- Next, find the solution to the wave equation
- Use method of "judicious guessing," guess:

$$V(x,t) = V_0 e^{-jkx} e^{j\omega t} = V_0 e^{-j\omega x/u} e^{j\omega t} = V_0 e^{-\gamma x} e^{j\omega t} \qquad jk = \gamma = \alpha + j\beta$$

- Where $k=\omega/u$ is the wavenumber in rad/m, ω is frequency in rad/s, and $u=v_p= (L_R C_R)^{-1/2}$ is the phase velocity in m/s
- Then, substituting into wave equation :

$$\frac{\partial^2 v(x,t)}{\partial x^2} = \frac{\partial^2 V_0 e^{-\gamma x} e^{j\omega t}}{\partial x^2} = L_R C_R \frac{\partial^2 v(x,t)}{\partial t^2} = L_R C_R \frac{\partial^2 V_0 e^{-\gamma x} e^{j\omega t}}{\partial t^2}$$

$$\Rightarrow \quad \gamma^2 V_0 e^{-\gamma x} e^{j\omega t} = -\omega^2 L_R C_R V_0 e^{-\gamma x} e^{j\omega t}$$

Note $\alpha=0$ for lossless line

- With propagation constant γ, and solution

$$\gamma = \pm j\omega\sqrt{L_R C_R} \quad \Rightarrow v(x,t) = V_0^+ e^{-j\omega\sqrt{L_R C_R}x} e^{j\omega t} + V_0^- e^{j\omega\sqrt{L_R C_R}x} e^{j\omega t}$$

- .

Phasor Notation

- We will use the most common engineering form of phasors
- Phasors are "shorthand" notation that have an "implied" $e^{j\omega t}$ term

if $\quad V(x,t) = |V_0| e^{j\theta} e^{-\gamma x} e^{j\omega t} = |V_0| e^{j\theta} e^{-\alpha x} e^{-j\beta x} e^{j\omega t} \quad$ for $\quad \gamma = \alpha + j\beta$

the shorthand phasor form of this would be:

$$V(x) = |V_0| e^{j\theta} e^{-\gamma x} = |V_0| e^{j\theta} e^{-\alpha x} e^{-j\beta x}$$

also, $|V_0| e^{j\theta}$ most commonly is denoted V_0, where V_0 is complex, so:

$$V(x) = V_0 e^{-\gamma x} = V_0 e^{-\alpha x} e^{-j\beta x}$$

$Note: \mathrm{Re}\{V(x,t)\} = \mathrm{Re}\{|V_0| e^{j\theta} e^{-\alpha x} e^{-j\beta x} e^{j\omega t}\} = |V_0| e^{-\alpha x} \cos(\omega t - \beta x + \theta)$

- BEWARE: physics literature often uses an "implied" $e^{-j\omega t}$ where there is a minus sign in the exponent
- Well-written journal articles should indicate which notation is being used by the authors ("time convention")

Beware of Different Notation or Conventions

- It is common in engineering literature to solve circuit equations, Maxwell's equations, and transmission line equations using phasors with the implied form $e^{j\omega t}$:

$$\text{Phasor notation with } e^{j\omega t} : E_0 e^{-\gamma z} e^{j\omega t} \text{ becomes phasor } E_0 e^{-\gamma z}$$
$$\text{and } \nabla \times H = J + \varepsilon \partial E / \partial t \quad \Rightarrow \quad \nabla \times H = J + j\omega\varepsilon E$$

- Beware of physics/optics literature (use of i vs j **may** forewarn) that often solves Maxwell's equations and transmission line equations using phasors of the form $e^{-j\omega t}$: Compare with previous

$$\text{Phasor notation with } e^{-i\omega t} : E_0 e^{ikz} e^{-i\omega t} \text{ becomes phasor } E_0 e^{ikz}$$
$$\text{and } \nabla \times H = J + \varepsilon \partial E_0 e^{ikz} e^{-i\omega t} / \partial t \quad \Rightarrow \quad \nabla \times H = J - i\omega\varepsilon E$$
$$\text{whereas an engneering text may use } E_0 e^{-jkz} e^{j\omega t}$$

Notably: Yariv of Cal Tech is "in the engineering camp"

Why Use Phasors and $e^{j\omega t}$?

- Beyond their power in providing uncluttered representation of complex functions, there is a deeper reason for using phasors and for interest in "$e^{j\omega t}$"
- That deeper reason is that the basic form "$A\,e^{-\gamma x}\,e^{j\omega t}$" provides a simple solution to many differential equations, such as Maxwell's equations and wave equations

$$\text{if } \frac{\partial^2 v(x,t)}{\partial x^2} = 9 \frac{\partial^2 v(x,t)}{\partial t^2} \text{ guess the solution is: } Ae^{-\gamma x} e^{j\omega t}$$

$$\Rightarrow \frac{\partial^2 Ae^{-\gamma x} e^{j\omega t}}{\partial x^2} = 9 \frac{\partial^2 Ae^{-\gamma x} e^{j\omega t}}{\partial t^2} \Rightarrow \gamma^2 \left(Ae^{-\gamma x} e^{j\omega t} \right) = -9\omega^2 \left(Ae^{-\gamma x} e^{j\omega t} \right)$$

$$\text{then } \gamma^2 = -9\omega^2 \text{ or } \gamma = j3\omega, \quad \pm\gamma \text{ are solutions, } \quad \text{so } Ae^{\pm j3\omega x} e^{j\omega t}$$

$$\text{in phasor notation, the solution is: } Ae^{\pm j3\omega x}$$

- Phasors are shorthand notation that have an "implied" $e^{j\omega t}$ term, and serve an important role in differential equations

Propagation Constant γ and Wavelength: λ

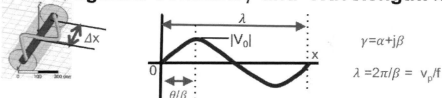

$$\gamma = \alpha + j\beta$$

$$\lambda = 2\pi/\beta = v_p/f$$

- For a traveling wave, wavelength λ is the physical length of one cycle in the transmission line, at an instant of time
- For a wave equation solution $\{|V_0|\, e^{j\theta}\, e^{j\omega t}\, e^{-\alpha x}\, e^{-j\beta x}\}$, the real wave would be $\mathrm{Re}\{|V_0|\, e^{j(\omega t+\theta)}\, e^{-\alpha x}\, e^{-j\beta x}\} = |V_0|\, e^{-\alpha x} \cos(\omega t+\theta-\beta x)$
- For $\alpha=0$, a "snapshot" at t=0 would look like the figure above
- Wavelength λ is the physical distance x that completes one cycle, corresponding to $\beta x = 2\pi$, or $\beta\lambda = 2\pi$
- Thus, wavelength in meters: $\lambda = 2\pi/\beta = 2\pi/(\omega/v_p) = v_p/f$
- Since propagation constant $\gamma = \alpha + j\beta$, the imaginary part of γ determines the wavelength

Frequency and Spatial Frequency

NOTE: axes are space and time

- For a traveling wave, in addition to the time-domain frequency in rad/s or Hz, there is a spatial frequency in rad/m
- For wave equation solution $\{|V_0|\, e^{j\theta}\, e^{j\omega t}\, e^{-\alpha x}\, e^{-j\beta x}\}$, the real wave is $\mathrm{Re}\{|V_0|\, e^{j(\omega t+\theta)}\, e^{-\alpha x}\, e^{-j\beta x}\} = \cos(\omega t+\theta-\beta x)$ for $V_0=1$, $\alpha=0$
- At any point x_0 on the line, we would observe a time-varying signal $\cos(\omega t+\theta-\beta x_0)$ with frequency $\omega=2\pi f=2\pi/T$ rad/s or f Hz
- At any time instant t_0, we would observe a voltage along the line $\cos(\omega t_0+\theta-\beta x)$ with spatial frequency $\beta=2\pi/\lambda$ rad/m
- This spatial frequency is commonly called wavenumber and $k=\beta$ in books using e^{-jkx} instead of $e^{-\gamma x}$ for lossless lines

Traveling Wave Phase Velocity

- The meaning of phase velocity can be illustrated by observing the wave in space at time 0 and time Δt

Time t=0, θ=0, α=0
$\text{Re}\{e^{j(\omega t+\theta)} e^{-\alpha x} e^{-j\beta x}\}$
$= \cos(\omega t - \beta x)$
$= \cos(-\beta x)$

Time t=Δt, θ=0, α=0
$\text{Re}\{e^{j(\omega t+\theta)} e^{-\alpha x} e^{-j\beta x}\}$
$= \cos(\omega t - \beta x)$
$= \cos(\omega \Delta t - \beta x)$

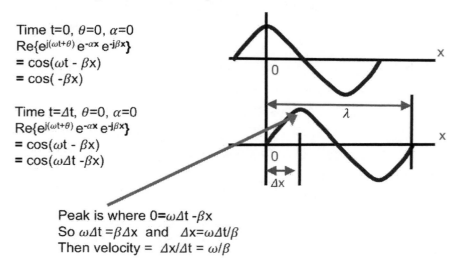

Peak is where $0 = \omega \Delta t - \beta x$
So $\omega \Delta t = \beta \Delta x$ and $\Delta x = \omega \Delta t / \beta$
Then velocity $= \Delta x/\Delta t = \omega/\beta$

Group Velocity

- Group velocity is the velocity of energy travel

group velocity $v_g = d\omega / d\beta = d\omega / dk$ is velocity of energy packet $\longrightarrow$ Envelope

consider: $\cos(a) + \cos(b) = 2\cos((a-b)/2)\cos((a+b)/2)$

for $\cos(\omega_1 t - \beta_1 x) + \cos(\omega_2 t - \beta_2 x) = 2\cos((\Delta\omega t - \Delta\beta x)/2)\cos((\Sigma\omega t - \Sigma\beta x)/2)$ At t=0, peak at x=0

where $\beta = \omega / v_p$, $\Delta\omega = \omega_1 - \omega_2$, and $\Delta\beta = \beta_1 - \beta_2$

at $t = 0$, the peak of the envelope is at $x = 0$, at $t = \Delta t$, the peak moves to Δx where

$(\Delta\omega \Delta t - \Delta\beta \Delta x)/2 = 0$ so $\Delta x = \Delta\omega \Delta t / \Delta\beta$

then, group velocity v_g is the envelope velocity

$v_g = \lim_{\Delta t \to 0} \frac{\Delta x}{\Delta t} = \lim_{\Delta\beta \to 0} \frac{\Delta\omega}{\Delta\beta} = \frac{d\omega}{d\beta}$

For the lossles LC line

$v_p = \omega / \beta = 1 / \sqrt{L_R C_R}$,

so, $\omega = \beta / \sqrt{L_R C_R}$, then

$v_g = \frac{d\omega}{d\beta} = 1 / \sqrt{L_R C_R} = v_p$

Example:
$v_g = v_p = 1$ m/s
$f_1 = 11$ Hz
$f_2 = 9$ Hz

Note: v_g does not always equal v_p

At t=0.1, peak at x=0.1

18

8

Lossy Transmission Line Model

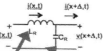

- Include resistance R (ohm/m) in series with L (H/m)
- Include conductance G (S/m) in parallel with C (F/m)

Lossy RLGC line propagation constant and phase velocity v_p

$$\gamma = \alpha + j\beta = \sqrt{(R_R + j\omega L_R)(G_R + j\omega C_R)}\ ,$$

where $V_0 e^{-\gamma x} e^{j\omega t} = V_0 e^{-(\alpha + j\beta)x} e^{j\omega t} = V_0 e^{-\alpha x} e^{-j\beta x} e^{j\omega t}$

and then $\beta = \text{Im}\{\gamma\} = \text{Im}\{\sqrt{(R_R + j\omega L_R)(G_R + j\omega C_R)}\}$

so wavelength $\lambda = v_p / f = 2\pi / \beta$

and phase velocity $v_p = \omega / \beta$

and attenuation constant α is

$$\alpha = \text{Re}\{\gamma\} = \text{Re}\{\sqrt{(R_R + j\omega L_R)(G_R + j\omega C_R)}\}$$

and characteristic impedance

$$Z_0 = \sqrt{(R + j\omega L)/(G + j\omega C)}$$

Impedance and Current For Lossy Line

- We solved for the voltage on the LC transmission line
- The current is

$$I(x,t) = \frac{V(x,t)}{Z_0} = \frac{V_0}{Z_0} e^{-\gamma x} e^{j\omega t} \quad where \quad Z_0 = \sqrt{\frac{(R_R + j\omega L_R)}{(G_R + j\omega C_R)}}$$

and for lossless right-handed LC lines $Z_0 = \sqrt{L_R / C_R}$

- Where Z_0 is the characteristic impedance of the line
- For a coaxial line, Z_0 and v_p are :

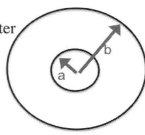

$$Z_0 = \frac{\sqrt{\mu/\varepsilon}\,\ln(b/a)}{2\pi} \quad \text{where a is inner radius, b is outer}$$

$$and \quad v_p = \frac{\omega}{\beta} = \lambda f = \frac{c}{\sqrt{\mu_r \varepsilon_r}}; \mu = \mu_r \mu_0, \varepsilon = \varepsilon_r \varepsilon_0$$

vacuum: $\mu = \mu_0 = 1257$ nH/m, $\varepsilon = \varepsilon_0 = 8.85$ pF/m

$c = 3 \times 10^8$ m/s

Example: Show I(x,t)=V(x,t)/Z$_0$

- Assume a lossless LC transmission line
- The Telegrapher's equations are

$$\frac{\partial v(x,t)}{\partial x} = -L_R \frac{\partial i(x,t)}{\partial t} \qquad \frac{\partial i(x,t)}{\partial x} = -C_R \frac{\partial v(x,t)}{\partial t}$$

$$\gamma = \alpha + j\beta = j\omega\sqrt{L_R C_R} \text{ , and } Z_0 = \sqrt{L_R / C_R} \text{ where:}$$

$$v(x,t) = V_0 e^{-\gamma x} e^{j\omega t} = V_0 e^{-j\beta x} e^{j\omega t} \text{ and } i(x,t) = \frac{V_0}{Z_0} e^{-\gamma x} e^{j\omega t} = \frac{V_0}{Z_0} e^{-j\beta x} e^{j\omega t}$$

- Substituting for v(x,t) and i(x,t)

$$\frac{\partial V_0 e^{-\gamma x} e^{j\omega t}}{\partial x} = -L_R \frac{V_0 e^{-\gamma x} e^{j\omega t} / Z_0}{\partial t} \quad and \quad \frac{\partial V_0 e^{-\gamma x} e^{j\omega t} / Z_0}{\partial x} = -C_R \frac{V_0 e^{-\gamma x} e^{j\omega t}}{\partial t}$$

$$\Rightarrow -\gamma V_0 e^{-\gamma x} e^{j\omega t} = -j\omega L_R V_0 e^{-\gamma x} e^{j\omega t} / Z_0 \quad and \quad -\gamma V_0 e^{-\gamma x} e^{j\omega t} / Z_0 = -j\omega C_R V_0 e^{-\gamma x} e^{j\omega t}$$

$$\Rightarrow -j\omega\sqrt{L_R C_R} = \frac{-j\omega L_R}{\sqrt{L_R / C_R}} \qquad and \qquad \frac{-j\omega\sqrt{L_R C_R}}{\sqrt{L_R / C_R}} = -j\omega C_R$$

$$\Rightarrow \boxed{-j\omega\sqrt{L_R C_R} = -j\omega\sqrt{L_R C_R}} \quad and \quad \boxed{-j\omega C_R = -j\omega C_R}$$

Coaxial Line Example

- For RG-58 coaxial transmission line
  - Polyethylene: $\varepsilon_r = 2.6$, b=0.131 inch, a=0.032 inch
  - C_R=94 pF/m

$\varepsilon_r = 2.6$, b=0.131in=3.33mm, a=0.032in=0.81mm

$\mu = \mu_0 = 1257$ nH/m, $\varepsilon = \varepsilon_r \varepsilon_0 = 2.6 \cdot 8.85$ pF/m = 23 pF/m

$$Z_0 = \frac{\sqrt{\mu/\varepsilon}\,\ln(b/a)}{2\pi} = \frac{\sqrt{\dfrac{1257\times10^{-9}\text{H}}{23\times10^{-12}\text{F}}}\,\ln\left(\dfrac{3.33\times10^{-3}\text{m}}{0.81\times10^{-3}\text{m}}\right)}{2\pi} = 52.5 \text{ ohms}$$

$$v_p = \frac{\omega}{\beta} = \lambda f = \frac{c}{\sqrt{\mu_r \varepsilon_r}}; \mu = \mu_r \mu_0, \varepsilon = \varepsilon_r \varepsilon_0$$

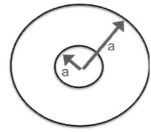

vacuum: $\mu = \mu_0 = 1257$ nH/m, $\varepsilon = \varepsilon_0 = 8.85$ pF/m

if $Z_0 = \sqrt{L_R / C_R}$, and $C_R = 94$ pF/m,

then $L_R = Z_0^2 C_R = (52.5)^2 \cdot (94\times10^{-12}) = 259$ nH/m

Lossy Coaxial Line Example

- For our RG-58 coaxial transmission line example

- Suppose R_R is 1 ohm/m and $G_R=0$

- What length has 3 dB attenuation at 1 GHz?

- 3 dB= 0.71 of voltage & half power

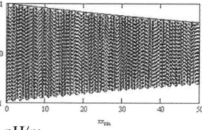

Recall for RG-58 : $C_R = 94$ pF/m, $L_R = 259$ nH/m

$$\gamma = \alpha + j\beta = \sqrt{(R_R + j\omega L_R)(G_R + j\omega C_R)}$$

$$= \sqrt{\left(1 + j\left(2\pi \times 10^9\right)\left(259 \times 10^{-9}\right)\right)\left(0 + \left(2\pi \times 10^9\right)\left(94 \times 10^{-12}\right)\right)} = 0.0095 + j31$$

where $V_0 e^{-\gamma x} e^{j\omega t} = V_0 e^{-\alpha x} e^{-j\beta x} e^{j\omega t} = V_0 e^{-0.0095x} e^{-j31x} e^{j2\pi \times 10^9 t}$

$$\text{Re}\left\{V_0 e^{-0.0095x} e^{-j31x} e^{j2\pi \times 10^9 t}\right\} = V_0 e^{-0.0095x} \cos\left(2\pi \times 10^9 t - 31x\right)$$

at $x = 36$ m, $e^{-0.0095x} = e^{-(0.0095)(36)} = e^{-0.34} = \dfrac{1}{\sqrt{2}}$

"Left-handed" transmission line wave equations

Metamaterial "Left-Handed" Transmission Lines

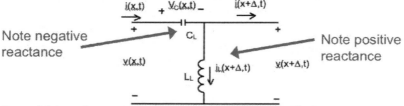

- Consider a lumped-element model of a "left-handed" line above
- The left-handed transmission line equations are

$$\frac{\partial^2 v(x,t)}{\partial x \partial t} = -\frac{1}{C_L} i(x,t) \qquad \frac{\partial^2 i(x,t)}{\partial x \partial t} = -\frac{1}{L_L} v(x,t)$$

- Units are L_L in H·m and C_L in F·m
- And substituting for $\partial i(x,t)/\partial t$ yields the "left-handed" <u>wave equation</u>:

$$\frac{\partial^4 v(x,t)}{\partial x^2 \partial t^2} = \frac{1}{L_L C_L} v(x,t)$$

Propagation in "Left-Handed" Transmission Lines

$$\frac{\partial^4 v(x,t)}{\partial x^2 \partial t^2} = \frac{1}{L_L C_L} v(x,t)$$

- Use "judicious guessing" and substituting in wave equation :

$$\frac{\partial^4 v(x,t)}{\partial x^2 \partial t^2} = \frac{\partial^4 V_0 e^{-\gamma x} e^{j\omega t}}{\partial x^2 \partial t^2} = \frac{1}{L_L C_L} v(x,t) = \frac{1}{L_L C_L} V_0 e^{-\gamma x} e^{j\omega t}$$

$$so: \quad -\gamma^2 \omega^2 V_0 e^{-\gamma x} e^{j\omega t} = \frac{1}{L_L C_L} V_0 e^{-\gamma x} e^{j\omega t} \quad \Rightarrow \gamma^2 = -1/\left(\omega^2 L_L C_L\right)$$

- Then, the solution is

$$\gamma = \alpha + j\beta = \frac{\pm j}{\omega \sqrt{L_L C_L}} \Rightarrow v(x,t) = V_0^+ e^{-jx/\left(\omega \sqrt{L_L C_L}\right)} e^{j\omega t} + V_0^- e^{jx/\left(\omega \sqrt{L_L C_L}\right)} e^{j\omega t}$$

Opposite directions
"Backward wave"

phase velocity $u = v_p = \omega / \beta = \omega^2 \sqrt{L_L C_L}$

group velocity $v_g = \dfrac{d\omega}{d\beta} = \dfrac{1}{d\beta / d\omega} = \dfrac{1}{d\left(\omega^{-1}/\sqrt{L_L C_L}\right)/ d\omega} = -\omega^2 \sqrt{L_L C_L}$

"Left-Handed" and "Right-Handed" Solutions

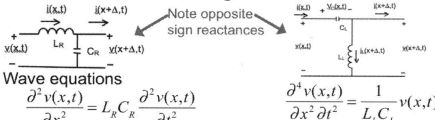

Note opposite sign reactances

- Wave equations

$$\frac{\partial^2 v(x,t)}{\partial x^2} = L_R C_R \frac{\partial^2 v(x,t)}{\partial t^2} \qquad \frac{\partial^4 v(x,t)}{\partial x^2 \partial t^2} = \frac{1}{L_L C_L} v(x,t)$$

- Then, plug in general solution for both wave equations

$$V(x,t) = V_0 e^{-jkx} e^{j\omega t} = V_0 e^{-j\omega x/u} e^{j\omega t} = V_0 e^{-\gamma x} e^{j\omega t} \qquad \gamma = \alpha + j\beta$$

- And get the phase velocities u_R and u_L and the solutions

$$v(x,t) = V_0^+ e^{-j\omega \sqrt{L_R C_R} x} e^{j\omega t} \qquad v(x,t) = V_0^+ e^{-jx/\left(\omega \sqrt{L_L C_L}\right)} e^{j\omega t}$$

Dispersive and "Backward wave"

$$v_p = \omega / \beta = 1/\sqrt{L_R C_R}, \qquad v_p = \omega / \beta = \omega^2 \sqrt{L_L C_L}$$

$$v_g = \frac{\partial \omega}{\partial \beta} = 1/\sqrt{L_R C_R} = v_p \qquad v_g = \frac{\partial \omega}{\partial \beta} = -\omega^2 \sqrt{L_L C_L} = -v_p$$

Right-handed solution   Left-handed solution

Comparison: "Left-Handed" and "Right-Handed" Velocities

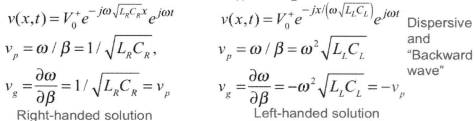

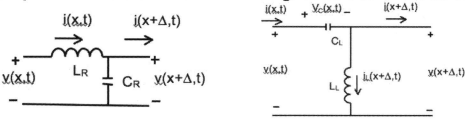

- Right-handed solution
- The velocities are constant
- The group and phase velocities are equal

- Left-handed solution
- The velocities change with frequency
- Group and phase velocities have opposite sign
- Called a "backward wave"

Note: group velocity gives the direction of energy flow

Analogy to Double-Negative Materials

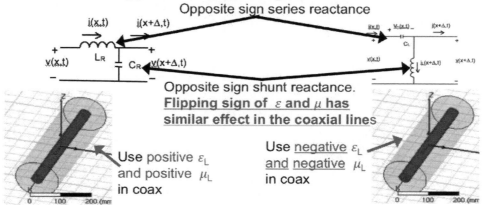

Opposite sign series reactance

Opposite sign shunt reactance.
Flipping sign of ε and μ has similar effect in the coaxial lines

Use positive ε_L and positive μ_L in coax

Use negative ε_L and negative μ_L in coax

- **Right-handed** LC line
- Analogous to a "**Normal**" transmission line
- Both permittivity ε_R and permeability μ_R are positive

29

- **Left-handed** LC line
- Analogous to a "**double-negative**" metamaterial (DNG)
- Both permittivity ε_L and permeability μ_L are negative

CRLH Structures

- A more common form for left-handed transmission lines is the composite right/left hand (CRLH) structure above
- This band-pass topology is left-handed at low frequencies and right-handed at high frequencies
- Analysis is along the same lines as before and would yield a more complicated solution
- For further details, see: Weldon, et al., "`Left-Handed Extensions of Maxwell's Equations for Metamaterials," IEEE SoutheastCon 2010 Proceedings, March 18-21, 2010.

Review of dB

Decibels

- Decibels, or dB, are $10 \log_{10}$ (power ratio)

  $10 \log_{10} (P / P_{ref})$
- Always power ratio!
- Typically relative to some reference

  1V, 1W, milliwatt, μV
- Common RF notation is dBx meaning dB relative to reference level "x"
- dBm = dB relative to 1 milliwatt
- Power in dBm units =

  $10 \log_{10}$ (power in milliwatts /1 milliwatt)
- Example: 0.0001 Watts =

  $10 \log_{10} (0.1 \text{ mW} / 1 \text{ mW}) = -10 \text{ dBm}$

Decibels 2

- Power is proportional to voltage squared or current squared
- So translating a voltage ratio into power ratio requires it to first be squared
- Useful properties of dB
  - Factors of 2 in power=3 dB
  - dB, factors of 10 in power=10 dB
- dB = 10 $\log_{10}$ (power ratio) =
  10 $\log_{10}$ [V^2 / $(V_{ref})^2$] =20 $\log_{10}$ (V / V_{ref})
- The 20 log comes from the square of the voltage
- Example: 10 V rms = 20 dBv
- What is 5 mW in dBm? In dBW?

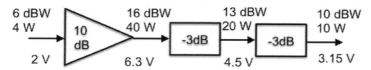

Common dB Units

- Frequency in radians/second = 2 π f, with frequency f in Hz
- Phase in radians
- dBm: dB relative to 1 milliwatt
- dBW: dB relative to 1 watt (0.1 watt = -10 dBW)
- dBV: dB relative to 1 volt rms (0.1 V = -20 dBV)
- dBc: dB relative to the carrier in a modulated signal
- dBi: dB relative to an isotropic antenna
- Bandwidth: 3 dB, zero-crossing, 90% energy
- Lowpass bandwidth: 0 to cutoff
- Bandpass bandwidth: cutoff to cutoff
- -3 dB is half-power, -6 dB is half-voltage
- 3 dB is double-power, 6 dB is double-voltage

16

2 WAVES, REFLECTION, AND S-PARAMETERS

The lecture notes in this chapter provide a brief introduction to waves, reflection, and S-parameters.

Waves and Reflection, Pulses

- In high speed systems, multiple reflections of pulses may be present on the wire connections of circuit boards
- These wires behave as transmission lines

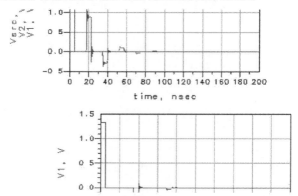

- Transmission lines, waves, reflections serve important role in high-frequency digital, RF, andmicrowave systems
- Optics and reflections are a good analogy [38]

Transmission Lines

- There are many different types of transmission line structures

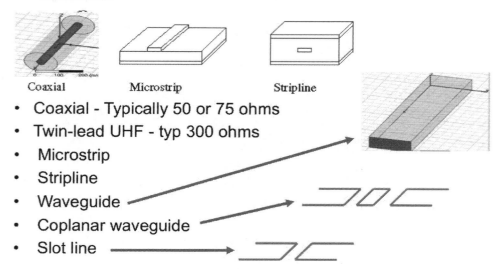

Coaxial   Microstrip   Stripline

- Coaxial - Typically 50 or 75 ohms
- Twin-lead UHF - typ 300 ohms
- Microstrip
- Stripline
- Waveguide
- Coplanar waveguide
- Slot line

Time Domain Reflections

- When time delay along a transmission line is large, we must consider wave properties of non-instantaneous transmission

- A good analogy is a laser and partial reflecting surfaces

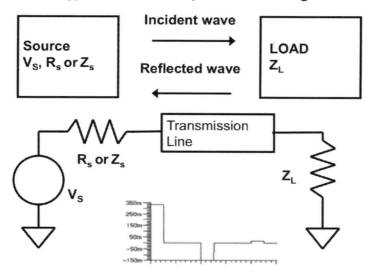

Reflection Coefficient

- It is easiest to analyze wave phenomena using the reflection coefficient (gamma):

$$\Gamma = \frac{Z_L - Z_S}{Z_L + Z_S}$$ where Z_L is load and Z_S is source impedance

Note: approach here may differ from some textbooks

At input interface, incident wave $V_+ = V_s / 2 = \dfrac{\text{open-circuit source voltage}}{2}$

At input interface, reflected wave $V_- = \Gamma V_+$

At input interface, total voltage $V_T = V_+ + V_- = (1 + \Gamma)V_s / 2 = \dfrac{Z_L V_s}{Z_L + Z_S}$

As a pulse makes multiple reflections, repeat the above process where

the source impedance becomes the line impedance,

and the load alternately becomes the generator and load

Simple Examples Without a Line

- Short circuit load

- Open circuit load

$$\Gamma = \frac{Z_L - Z_S}{Z_L + Z_S} = \frac{0 - Z_S}{0 + Z_S} = -1$$

incident wave $V_{inc} = V_s / 2$

reflected $V_r = \Gamma V_{inc} = -V_{inc} = -V_s / 2$

total voltage:

$$V_T = V_{inc} + V_r = (1 + \Gamma)V_s / 2 = 0$$

As expected,

0 volts across the short circuit

42

$$\Gamma = \frac{Z_L - Z_S}{Z_L + Z_S} = \frac{\infty - Z_S}{\infty + Z_S} = 1$$

incident wave $V_{inc} = V_s / 2$

reflected $V_r = \Gamma V_{inc} = V_{inc} = V_s / 2$

total voltage:

$$V_T = V_{inc} + V_r = (1 + \Gamma)V_s / 2 = V_s$$

As expected,

all volts across the open circuit

Simple Examples Without a Line

- Resistive matched load

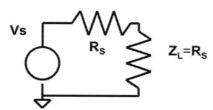

$$\Gamma = \frac{Z_L - Z_S}{Z_L + Z_S} = \frac{R_S - R_S}{R_S + R_S} = 0$$

incident wave $V_{inc} = V_s / 2$

reflected $V_r = \Gamma V_{inc} = 0$

total voltage:

$$V_T = V_+ + V_{inc} = (1 + \Gamma)V_s / 2 = V_s / 2$$

As expected,

the max power match voltage equals $V_s / 2$

Simple Examples Without a Line

- Conjugate-matched load

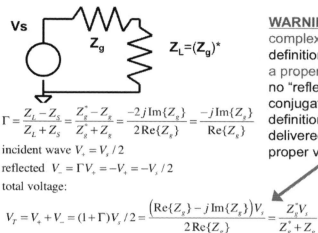

WARNING: when Z_S is complex, clearly this definition of Γ does NOT give a proper sense of there being no "reflected power" under conjugate match, where by definition maximum power is delivered. However, the proper voltage is computed.

$$\Gamma = \frac{Z_L - Z_S}{Z_L + Z_S} = \frac{Z_g^* - Z_g}{Z_g^* + Z_g} = \frac{-2j\,\mathrm{Im}\{Z_g\}}{2\,\mathrm{Re}\{Z_g\}} = \frac{-j\,\mathrm{Im}\{Z_g\}}{\mathrm{Re}\{Z_g\}}$$

incident wave $V_+ = V_s / 2$

reflected $V_- = \Gamma V_+ = -V_+ = -V_s / 2$

total voltage:

$$V_T = V_+ + V_- = (1+\Gamma)V_s / 2 = \frac{\left(\mathrm{Re}\{Z_g\} - j\,\mathrm{Im}\{Z_g\}\right)V_s}{2\,\mathrm{Re}\{Z_g\}} = \frac{Z_g^* V_s}{Z_g^* + Z_g}$$

As expected,

the conjugate match voltage equals that of the voltage divider comprised of Z_g^* and Z_g

See classic book by Collin, for a different definition of Γ

Multiple Reflections in a Lossless Transmission Line

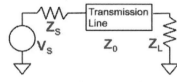

- Reflections can occur at discontinuities
- Reflection coefficient is computed at every point of discontinuity/transition
- Here, discontinuities are at both ends of the transmission line

incident wave here: $V_s / 2$

$$\Gamma_1 = \frac{Z_0 - Z_S}{Z_0 + Z_S}$$

$$V_{R1} = \Gamma_1 V_{inc} = \frac{Z_0 - Z_S}{Z_0 + Z_S} \frac{V_s}{2}$$

$$V_{I1} = (1+\Gamma_1)V_{inc} = Z_0 V_s / (Z_0 + Z_S)$$

$$\Gamma_3 = (Z_S - Z_0)/(Z_S + Z_0) = -\Gamma_1$$

$$V_{I2} = (1+\Gamma_3)V_{inc}$$

$$= (1-\Gamma_1)V_{inc}$$

$$= \Gamma_2(1-\Gamma_1^2)V_s / 2$$

$$\Gamma_3 V_{inc} = -\Gamma_1\Gamma_2(1+\Gamma_1)V_s / 2$$

$$\Gamma_2 = (Z_L - Z_0)/(Z_L + Z_0)$$

$$V_{L1} = (1+\Gamma_2)V_{inc}$$

$$= (1+\Gamma_2)\left((1+\Gamma_1)V_s / 2\right)$$

$$\Gamma_2 V_{inc} = \Gamma_2\left((1+\Gamma_1)V_s / 2\right)$$

$$\Gamma_4 = \Gamma_2 = (Z_L - Z_0)/(Z_L + Z_0)$$

$$V_{L2} = (1+\Gamma_4)V_{inc}$$

$$= -\Gamma_1\Gamma_2(1+\Gamma_2)(1+\Gamma_1)\frac{V_s}{2}$$

$$\Gamma_4 V_{inc} = -\Gamma_1\Gamma_2^2(1+\Gamma_1)\frac{V_s}{2}$$

etc.

Short Pulse Example

$Z_s=R_s= 100$, $Z_0 = 50$, $Z_L=R_L= 25$
Transmission line delay = 100 ns
Open-circuit source voltage V_S=1 V

$$\Gamma_1 = \frac{Z_0 - R_s}{Z_0 + R_s} = \frac{50 - 100}{50 + 100} = -0.33$$

first reflection $\Gamma_1 V_{inc} = -0.33 V_s / 2 = -0.165$

$$V_{I1} = (1 + \Gamma_1) V_S / 2 = (1 - 0.33)/2 = 0.33$$

$$\Gamma_2 = \frac{R_L - Z_0}{R_L + Z_0} = \frac{25 - 50}{25 + 50} = -0.33$$

$$\Gamma_2 V_{inc} = -0.33 V_{+1} = -0.11$$

$$V_{L1} = (1 + \Gamma_2) V_{inc} = (1 + -0.33)(0.33) = 0.22$$

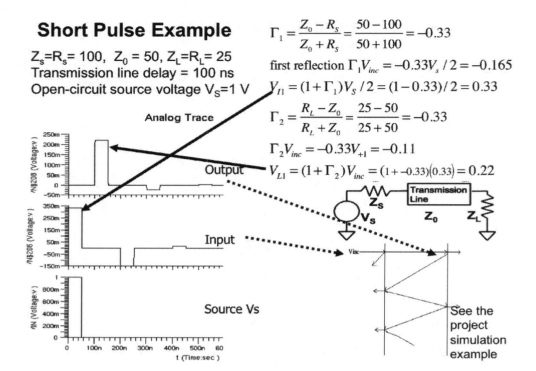

See the project simulation example

S-Parameters

Reflection Coefficient Summary

- Incident wave V_i
- Define V_i as $V_S/2$
- Reflected wave: $V_r = \Gamma V_i$
- Reflection coefficient Γ
  $$\Gamma = (Z_L - Z_s) / (Z_L + Z_s)$$
- Total voltage $V_t = V_i + V_r$ = transmitted wave
- Z_L is load impedance, Z_s is source impedance
- If source is a transmission line, $Z_s = Z_0$

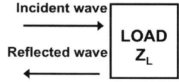

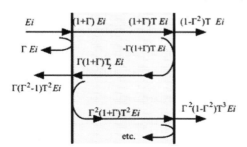

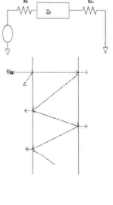

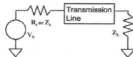

Return Loss and VSWR

- **Return loss** = Ratio of reflected power to incident power:
  $$P_r / P_i = |\Gamma|^2$$

  Note minus sign, so +10 dB means return is 10 dB below incident

  o Return loss measured in dB:
  $$\text{return loss} = -10 \log_{10}(P_r / P_i) = -10 \log_{10}(|\Gamma|^2)$$

  o 20 dB return loss: 1% power reflected, 99% delivered to load

- VSWR is Voltage Standing Wave Ratio
  o Then, VSWR is defined as:
  $$\text{VSWR} = (1 + |\Gamma|) / (1 - |\Gamma|)$$

  o Reflection produces a standing wave
    Over a long length of line, the maximum amplitude will be
    $V_{max} = 1 + |\Gamma|$ and minimum amplitude $V_{min} = 1 - |\Gamma|$

  o The max and min can be measured along the line

2-Port Network

- A 2-port network is a device with 2 ports, typically input and output
- Examples include amplifiers, filters, attenuators
- An input signal can be applied from an external source to either port
- Amplifiers: forward gives gain, reverse is isolation
- Typically attenuators are reciprocal (symmetric behavior)
- To fully characterize: both directions must be measured
- Each port has an incident wave and reflected wave

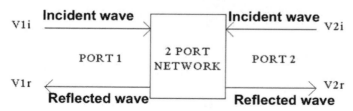

S-Parameters

- S-parameters (scattering parameters) used to characterize devices
- Define normalized parameters
    - $a_1 = V_{1i} / (Z_0)^{1/2}$, incident wave to port 1
    - $a_2 = V_{2i} / (Z_0)^{1/2}$, incident wave to port 2
    - $b_1 = V_{1r} / (Z_0)^{1/2}$, output wave from port 1
    - $b_2 = V_{2r} / (Z_0)^{1/2}$, output wave from port 2

      Where we assume V is peak

    Then: $0.5|a_1|^2$ = incident power to port 1, etc.

    The S-parameters are defined as:
    - $b_1 = S_{11} a_1 + S_{12} a_2$, output wave of port 1
    - $b_2 = S_{21} a_1 + S_{22} a_2$, output wave of port 2

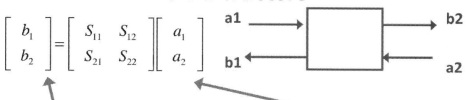

$$52 \quad \begin{bmatrix} b_1 \\ b_2 \end{bmatrix} = \begin{bmatrix} S_{11} & S_{12} \\ S_{21} & S_{22} \end{bmatrix} \begin{bmatrix} a_1 \\ a_2 \end{bmatrix}$$

S-Parameters

$$\begin{bmatrix} b_1 \\ b_2 \end{bmatrix} = \begin{bmatrix} S_{11} & S_{12} \\ S_{21} & S_{22} \end{bmatrix} \begin{bmatrix} a_1 \\ a_2 \end{bmatrix}$$

- The left vector is the output wave of the 2 ports, the right vector is the input wave to the two ports
- For a reciprocal 2-port, the matrix is symmetric, i.e., $S_{21} = S_{12}$
- All passive RLC 2-ports are reciprocal
- Can also be represented with an S-parameter flowgraph diagram:

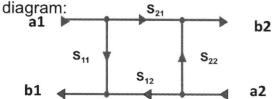

Measurement of S-parameters

- If $a_2=0$, then:
  - $b_1 = S_{11}\, a_1 + S_{12}\, a_2 = S_{11}\, a_1$, output wave of port 1
  - $b_2 = S_{21}\, a_1 + S_{22}\, a_2 = S_{21}\, a_1$, output wave of port 2
- Note: setting $a_2=0$ <u>does not change the load</u> at port 2, the load impedance on port 2 remains Z_0
- Then, with $a_2=0$ and a test signal (input wave) at port 1:

$$S_{11} = \left.\frac{b_1}{a_1}\right|_{a_2=0} = \left.\frac{V_{1r}/\sqrt{Z_0}}{V_{1i}/\sqrt{Z_0}}\right|_{a_2=0} = \left.\frac{V_{1r}}{V_{1i}}\right|_{a_2=0}$$

- S_{11} is Γ for port 1 when $a_2=0$ and the load on port 2 is Z_0
- Must terminate port 2 to set $a_2=0$ (no reflection from load)

$$\begin{bmatrix} b_1 \\ b_2 \end{bmatrix} = \begin{bmatrix} S_{11} & S_{12} \\ S_{21} & S_{22} \end{bmatrix} \begin{bmatrix} a_1 \\ a_2 \end{bmatrix}$$

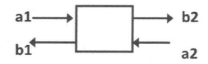

Measurement of S-parameters

- The second S-parameter measured with $a_2=0$ is

$$S_{21} = \left.\frac{b_2}{a_1}\right|_{a_2=0} = \left.\frac{V_{2r}/\sqrt{Z_0}}{V_{1i}/\sqrt{Z_0}}\right|_{a_2=0} = \left.\frac{V_{2r}}{V_{1i}}\right|_{a_2=0}$$

- S_{21} gives the output of port 2 for an input at port 1 when $a_2=0$ and the load on port 2 is Z_0
- Finally, S_{22} and S_{12} are determined as:

$$S_{22} = \left.\frac{b_2}{a_2}\right|_{a_1=0} \qquad S_{12} = \left.\frac{b_1}{a_2}\right|_{a_1=0}$$

- Note: S-parameter S_{21} is gain

$$\begin{bmatrix} b_1 \\ b_2 \end{bmatrix} = \begin{bmatrix} S_{11} & S_{12} \\ S_{21} & S_{22} \end{bmatrix} \begin{bmatrix} a_1 \\ a_2 \end{bmatrix}$$

Measurement of S-parameters

- In our notation |V| is peak voltage, and $a_1 = V_{1i} / (Z_0)^{1/2}$
- Then, $0.5|a_1|^2$ and $0.5|b_1|^2$ are incident and reflected power at port 1 when port 2 is terminated in Z_0, so:

$$|S_{11}|^2 = \left.\frac{|b_1|^2 / 2}{|a_1|^2 / 2}\right|_{a_2=0} = \left.\frac{P_{1r}}{P_{1i}}\right|_{a_2=0} = \left.\frac{\text{reflected power}}{\text{incident power}}\right|_{a_2=0}$$

- Similarly,

$$|S_{21}|^2 = \left.\frac{|b_2|^2 / 2}{|a_1|^2 / 2}\right|_{a_2=0} = \left.\frac{P_{2r}}{P_{1i}}\right|_{a_2=0} = \left.\frac{\text{output power}}{\text{incident power}}\right|_{a_2=0}$$

$$\begin{bmatrix} b_1 \\ b_2 \end{bmatrix} = \begin{bmatrix} S_{11} & S_{12} \\ S_{21} & S_{22} \end{bmatrix} \begin{bmatrix} a_1 \\ a_2 \end{bmatrix}$$

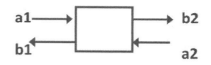

S-parameters in dB, with Phase

- S-parameters are often plotted in dB and with phase plots
- Again, in our case |V| is peak voltage, and $a_1 = V_{1i} / (Z_0)^{1/2}$
- Then, $0.5|a_1|^2$ and $0.5|b_1|^2$ are incident and reflected power at port 1 when port 2 is terminated in Z_0, so:

$$10\log_{10}\left(|S_{11}|^2\right) = \left.\frac{\text{reflected power}}{\text{incident power}}\right|_{a_2=0} \text{ in dB} \quad and \quad 10\log_{10}\left(|S_{21}|^2\right) = \left.\frac{\text{output power}}{\text{incident power}}\right|_{a_2=0} \text{ in dB}$$

- For example:

    if $S_{11} = 0.2e^{-j\pi/4}$ then it can also be written as $S_{11} = -14$ dB $\angle -45°$

    if $S_{21} = 8e^{j\pi/2}$ then it can also be written as $S_{21} = 18$ dB $\angle 90°$

- Example plots:

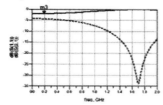
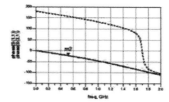

Example: S-parameters

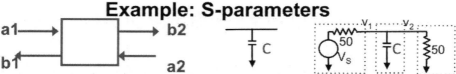

- Example: in a Z_0=50-ohm system, find the S-parameters of the circuit above, consisting of a capacitor to ground

$$S_{11} = \left.\frac{b_1}{a_1}\right|_{a_2=0} = \Gamma_1 = \frac{v_{ref}}{v_{inc}} = \frac{Z-Z_0}{Z+Z_0} = \frac{50/(1+j50\omega C)-50}{50/(1+j50\omega C)+50} = \frac{-j2500\omega C}{100+j2500\omega C}$$

$$S_{22} = \left.\frac{b_2}{a_2}\right|_{a_1=0} = \Gamma_2 = \frac{Z-Z_0}{Z+Z_0} = \frac{50/(1+j50\omega C)-50}{50/(1+j50\omega C)+50} = \frac{-j2500\omega C}{100+j2500\omega C}$$

$$S_{21} = \left.\frac{b_2}{a_1}\right|_{a2=0} = \frac{v_2}{v_{inc}} = \frac{v_2}{V_S/2} = \frac{\left(\dfrac{50/(1+j50\omega C)}{50+50/(1+j50\omega C)}V_s\right)}{V_S/2} = \frac{100}{100+j2500\omega C}$$

$$S_{12} = S_{21}$$

$$\begin{bmatrix} b_1 \\ b_2 \end{bmatrix} = \begin{bmatrix} \dfrac{-j2500\omega C}{100+j2500\omega C} & \dfrac{100}{100+j2500\omega C} \\ \dfrac{100}{100+j2500\omega C} & \dfrac{-j2500\omega C}{100+j2500\omega C} \end{bmatrix} \begin{bmatrix} a_1 \\ a_2 \end{bmatrix}$$

Example: S-parameters

- Example: in a 50-ohm system, find the S-parameters of the circuit above

$$S_{11} = \left.\frac{b_1}{a_1}\right|_{a_2=0} = \Gamma_1 = \frac{v_{ref}}{v_{inc}} = \frac{Z-Z_0}{Z+Z_0} = \frac{75-50}{75+50} = 0.2$$

$$S_{22} = \left.\frac{b_2}{a_2}\right|_{a_1=0} = \Gamma_2 = \frac{Z-Z_0}{Z+Z_0} = \frac{50(100)/150-50}{50(100)/150+50} = -0.2 = 0.2e^{-j\pi}$$

$$S_{21} = \left.\frac{b_2}{a_1}\right|_{a2=0} = \frac{v_2}{v_{inc}} = \frac{v_2}{V_S/2} = \frac{\left(\dfrac{25}{125}V_s\right)}{V_S/2} = 0.4$$

$$S_{12} = \left.\frac{b_2}{a_1}\right|_{a1=0} = \frac{v_1}{v_{inc}} = \frac{v_1}{V_S/2} = \frac{\left(\dfrac{33}{83}\dfrac{50}{100}V_s\right)}{V_S/2} = 0.4 \qquad \begin{bmatrix} b_1 \\ b_2 \end{bmatrix} = \begin{bmatrix} 0.2 & 0.4 \\ 0.4 & 0.2e^{-j\pi} \end{bmatrix} \begin{bmatrix} a_1 \\ a_2 \end{bmatrix}$$

Example: S-parameters

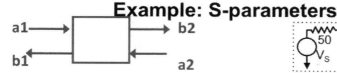

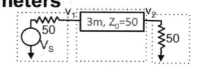

- Example: in a 50-ohm system, find the S-parameters of a 3 m long 50 ohm transmission line with $\gamma = \alpha + j\beta = j2\pi/\lambda = j\omega/v_p$

since no reflection for 50 ohm cable in 5-ohm system, $S_{11} = S_{22} = 0$

$$S_{21} = \frac{b_2}{a_1}\bigg|_{a2=0} = \frac{v_2}{v_{inc}} = \frac{v_2}{V_S/2} = \frac{(V_S/2)e^{-\gamma x}}{V_S/2} = e^{-j\omega x/v_p} = e^{-j3\omega/v_p}$$

$$S_{12} = S_{21}$$

- So:

$$\begin{bmatrix} b_1 \\ b_2 \end{bmatrix} = \begin{bmatrix} 0 & e^{-j3\omega/v_p} \\ e^{-j3\omega/v_p} & 0 \end{bmatrix} \begin{bmatrix} a_1 \\ a_2 \end{bmatrix}$$

Frequency Dependence of S-parameters

- The example of the S-parameters of a capacitor were frequency dependent (some parameters included "ω")

$$\begin{bmatrix} b_1 \\ b_2 \end{bmatrix} = \begin{bmatrix} \frac{-j2500\omega C}{100 + j2500\omega C} & \frac{100}{100 + j2500\omega C} \\ \frac{100}{100 + j2500\omega C} & \frac{-j2500\omega C}{100 + j2500\omega C} \end{bmatrix} \begin{bmatrix} a_1 \\ a_2 \end{bmatrix}$$

- Whereas the S-parameter matrix for the resistive network was not frequency dependent (no parameters include "ω")

$$\begin{bmatrix} b_1 \\ b_2 \end{bmatrix} = \begin{bmatrix} 0.2 & 0.4 \\ 0.4 & 0.2e^{-j\pi} \end{bmatrix} \begin{bmatrix} a_1 \\ a_2 \end{bmatrix}$$

- **NOTE**: In our earlier time-domain reflection coefficients Γ for pulses, we assumed Γ was not frequency-dependent
- **NOTE**: A frequency-dependent Γ would be expected to distort a time-domain pulse, just as a low-pass filter would be expected to distort a square pulse

Reciprocal Devices and S-parameters

- A reciprocal device has a symmetric S-parameter matrix
- A device is reciprocal if it has a symmetric S-parameter matrix
- An S-parameter matrix is symmetric if $S_{ij} = S_{ji}$ for all i and j
- For a 2-port, a device is reciprocal if $S_{21} = S_{12}$

$$\begin{bmatrix} b_1 \\ b_2 \end{bmatrix} = \begin{bmatrix} S_{11} & S_{12} \\ S_{21} & S_{22} \end{bmatrix} \begin{bmatrix} a_1 \\ a_2 \end{bmatrix}$$

- Passive RLC circuits are reciprocal
  - Our 2 passive examples were reciprocal
- Amplifiers are not typically reciprocal

Load and Source Effects on Reflection Coefficients

- Consider multiple reflections and reflection coefficient:

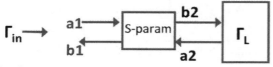

$$\Gamma_{in} = S_{11} + \frac{S_{12}S_{21}\Gamma_L}{1 - S_{22}\Gamma_L}$$

incident wave

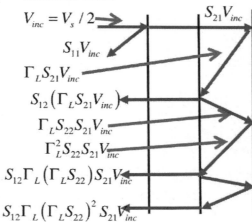

$$V_{b1tot} = \left(S_{11} + S_{12}\Gamma_L S_{21} \sum_{n=0}^{\infty} (S_{22}\Gamma_L)^n \right) V_{inc}$$

$$= \left(S_{11} + \frac{S_{21}S_{12}\Gamma_L}{1 - S_{22}\Gamma_L} \right) V_{inc} = \Gamma_{in} V_{inc}$$

$$V_{b2tot} = \left(\sum_{n=0}^{\infty} (S_{22}\Gamma_L)^n \right) S_{21} V_{inc}$$

$$= \frac{S_{21}}{1 - S_{22}\Gamma_L} V_{inc}$$

Load and Source Effects on Reflection Coefficients

- S-parameters are measured with Z_0 source and terminations
- Input and output reflection coefficients will change for other source/load conditions
- The reflection at the input Γ_{in} with load Γ_L is:

$$\Gamma_{in} = S_{11} + \frac{S_{21}S_{12}\Gamma_L}{1 - S_{22}\Gamma_L}$$

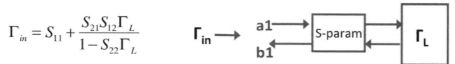

- The reflection at the output Γ_{out} with source Γ_S is:

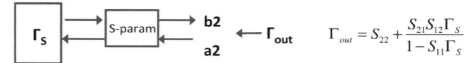

$$\Gamma_{out} = S_{22} + \frac{S_{21}S_{12}\Gamma_S}{1 - S_{11}\Gamma_S}$$

Example: Load and Source Effects

- For an amplifier with the following S-parameters that were measured with $Z_0 = 50$, find the input reflection coefficient when the amplifier load is 25 ohms
- Amplifier S-parameters:

$$\begin{bmatrix} S_{11} & S_{12} \\ S_{21} & S_{22} \end{bmatrix} = \begin{bmatrix} 0.2 & 0.1e^{-j\pi} \\ 20e^{-j\pi/2} & -0.25 \end{bmatrix}$$

- Solution:

$$\Gamma_L = (Z_L - Z_0)/(Z_L + Z_0) = (25 - 50)/(25 + 50) = -0.33 = 0.33e^{-j\pi}$$

$$\Gamma_{in} = S_{11} + \frac{S_{21}S_{12}\Gamma_L}{1 - S_{22}\Gamma_L} = 0.2 + \frac{(20e^{-j\pi/2})(0.1e^{-j\pi})(0.33e^{-j\pi})}{1 - (0.25e^{-j\pi})(0.33e^{-j\pi})}$$

$$= 0.2 + \frac{0.66e^{-j\pi/2}}{1 - 0.0825} = 0.2 - j0.73 = 0.75e^{-j1.3} = -2.4 \ dB \ \angle -75°$$

Source and Load Effects on Gain

- Transducer gain, g_t
- g_t= power delivered to load divided by power available from source:

$$g_t = \frac{\text{power delivered to load}}{\text{maximum power available from source}}$$

- Can be computed from measured s-parameters

$$g_t = \frac{|S_{21}|^2 (1-|\Gamma_S|^2)(1-|\Gamma_L|^2)}{|(1-S_{11}\Gamma_S)(1-S_{22}\Gamma_L)-S_{21}S_{12}\Gamma_S\Gamma_L|^2}$$

- What is transducer gain if source and load impedance = Z_0 ?
  - Answer: $|S_{21}|^2$
  - In dB: $10 \log1_0(|S_{21}|^2)$ dB

3 SMITH CHART AND IMPEDANCE MATCHING

The lecture notes in this chapter discuss the Smith chart and impedance matching.

Smith Chart and Impedance Matching

Smith Chart Applications

- Instrument readout for measurements
- Display RF simulation results
- Device data sheets (data given as Smith chart)
- Design aid
  - o Impedance matching
  - o Transmission line effects

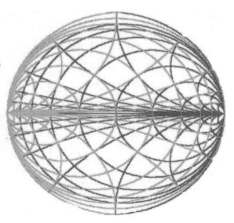

Smith Chart: Impedances and Admittances

- Impedances $Z = R + jX$ are normalized (resistance R and reactance X):
  - $Z_n = Z / Z_0$
- Unless otherwise noted: assume $Z_0 = 50$
- So,

  If $Z = 100 + j\,50$, then $Z_n = 2 + j\,1$
- Red lines are impedance, green are admittance on Smith chart
- Admittances $Y = G + jB$ are also normalized (conductance G and susceptance B):

  $Y_n = 1 / Z_n = Y / Y_0 = Z_0 / Z$
- Same example, if $Y = 0.008 - j\,0.004$,

  $Y_n = (0.008 - j\,0.004) / 0.02 = 0.4 - j\,0.2$
- Z_n and Y_n above are same point on Smith chart
- (see red circle on next slide)

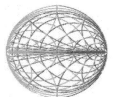

$$Y_n = 0.4 - j\,0.2$$
$$Z_n = 2 + j\,1$$

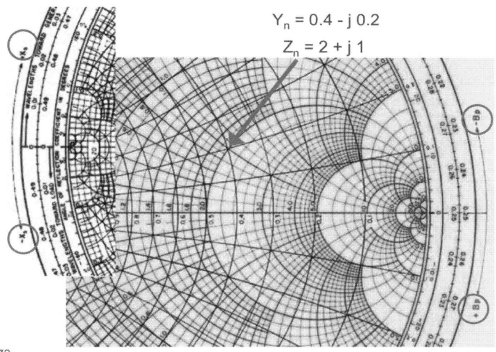

73

Smith Chart Orientation

- Pure real impedances along horizontal axis
- Inductive impedances in top half
- Pure inductance along upper perimeter
- Capacitive impedances in lower half
- Pure capacitance along lower perimeter
- Short circuit at left on horizontal axis
- Open circuit at right on horizontal axis
- Red lines – impedances (Z)
- Green lines – admittances (Y)

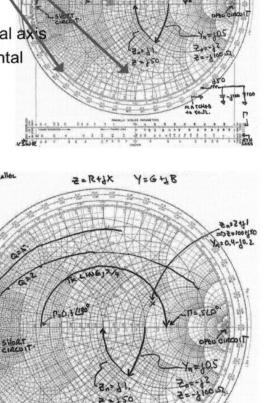

Smith Chart Orientation

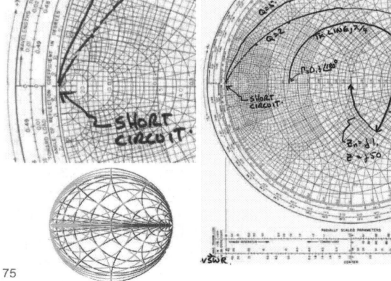

75

Smith Reflection Coefficient Γ angles

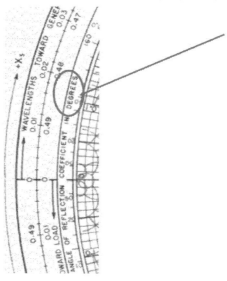

76

Smith Chart: Scales

- Scales for reflection coeff. Γ, return loss, VSWR
- Previous example: $Z = 100 + j\,50$, $Z_n = 2 + j\,1$

$\Gamma = (Z - Z_0) / (Z + Z_0)$

$= (100 + j\,50 - 50) / (100 + j\,50 + 50)$

$= 0.45 \quad \angle 27°$

-Use compass to read Γ on vol. refl. coeff. scale

-Read angle of Γ off perimeter of Smith chart

-Read return loss ($-10\log_{10}|\Gamma|^2 = 6.9$ dB) from scale

-Read VSWR $= (1 + |\Gamma_{load}|) / (1 - |\Gamma_{load}|) = 2.6$ (voltage ratio)

|Γ|

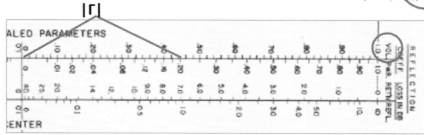

Smith Scales

$Z=100 + j\,50,\ Z_n = 2 + j\,1$
$\Gamma = (Z-Z_0) / (Z+Z_0)= 0.45\ \angle 27°$

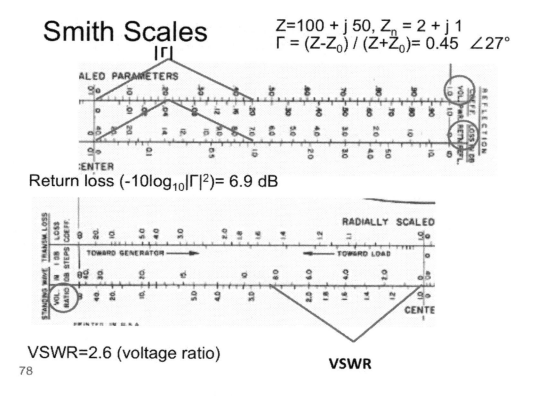

Return loss $(-10\log_{10}|\Gamma|^2)= 6.9$ dB

VSWR=2.6 (voltage ratio)

78

VSWR

Smith Chart Scales

- Example 2:
- Let Z=100 and Z_0=50, then $Z_n = 2 + j\,0$

  $\Gamma = (Z-Z_0) / (Z+Z_0)$

  $\quad = (100 - 50) / (100 + 50)$

  $\quad = 0.33\ \angle 0°$

  -Use compass to read Γ on vol. refl. coeff. scale

  -Read angle of Γ off perimeter of Smith chart

  -Read return loss $(-10\log_{10}|\Gamma|^2= 9.6$ dB) from scale

  -Read VSWR=$(1+ |\Gamma_{load}|) / (1- |\Gamma_{load}|) = 2$ (volt ratio)

- Note: better return losses lie within circles centered at Z_n=1

Series-to-Parallel Transformations

- Check this example on the Smith chart
- At a given frequency, an impedance can be represented as a series R_s and X_s or a parallel R_p (=$1/G_p$) and X_p (= $-1/B_p$)
- $Z = R_s + j\, X_s$ and $Y = G_p + j\, B_p$
- To convert:

$$R_p = R_s[1+(X_s/R_s)^2] \qquad X_p = X_s[1+(R_s/X_s)^2]$$

$$R_s = \frac{R_p}{[1+(R_p/X_p)^2]} \qquad X_s = \frac{X_p}{[1+(X_p/R_p)^2]}$$

- In our example $Z = 100 + j\,50 = R_s + j\,X_s$
- $R_p = 100[1+(50/100)^2] = 125\ \Omega$ ($G_p = 1/R_p = 0.008$ S)
- $X_p = 50[1+(100/50)^2] = 250\ \Omega$ ($B_p = -1/X_p = -0.004$ S)

$R_s = 100$

$Z = R_s + jX_s$ $jX_s = j50$ ➡ $Z = R_s + jX_s$ $R_p = 125$ $X_p = j250$

Impedance Matching

Smith Chart: Impedance Matching

- Series L and C travel on red, Parallel L and C travel on green
- Reactance/Admittance of series/parallel element is difference from start/stop
- Example match into 100 ohms:
  - Green (parallel C): travel down $Y_n = +j0.5$ S,
    $Y = Y_n Y_0 = (j0.5)(0.02) = j0.01$
    $Z = Z_n Z_0 = (1/Y_n) Z_0 = (-j2)(50) = -j100 = 1/Y$
  - Red (series L): travel up $Z_n = +j1\ \Omega$
    $Z = Z_n Z_0 = (j1)(50) = j50$
- So: $X_c = 1/Y_c = -j100\ \Omega = 1/(j\omega C)$, $X_L = j50\ \Omega = j\omega l$
- If at 1 GHz, C = 1.6 pF, L = 8 nH
- $Q = X/R : Q \approx F_0/B_{3dB}$

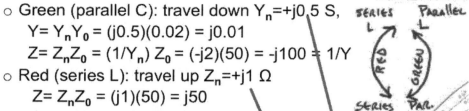

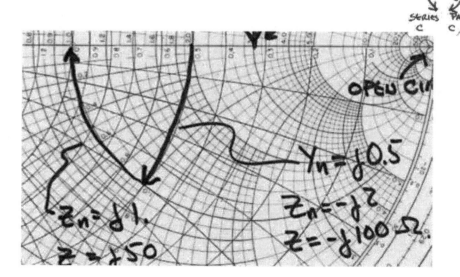

Smith Match Example

Alternative Match

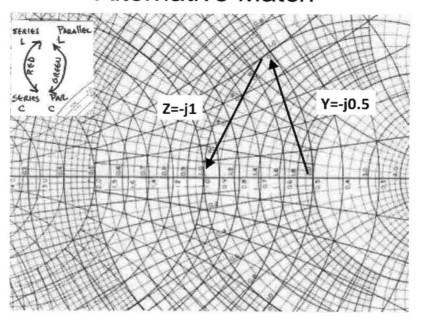

84

There are many
possibilities
for an impedance
matching network

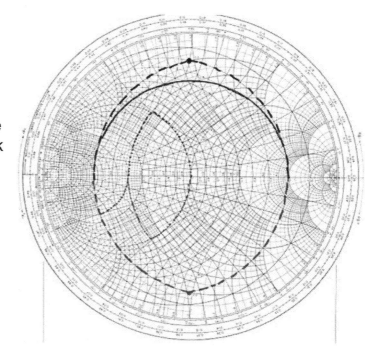

85

Smith Chart: Transmission Line Matching

- When a load is at the end of a transmission line,
- To find the impedance looking down the line at the l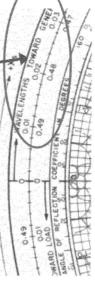
  - Travel in circle on the Smith chart, clockwise sta from load and proceeding "toward generator"
- Z_0 of trans. line MUST be center of chart
- Example (note: we are not using 50 ohm line):
  $\lambda/4$ Line with Z_0=150 and 50 ohm load at far end
- Starting point at load: Z_n=50/150=0.33 (normalized)
- Here, $\Gamma_{load} = (Z_L-Z_0) / (Z_L+Z_0)$=0.5 $\angle 180°$
  -Use compass to read Γ_{load} on vol. refl. coeff. scale
  -Read angle of Γ_{load} off perimeter of Smith chart
  -Read return loss ($-10\log_{10}|\Gamma_{load}|^2$= 6 dB) from scale
  -Read VSWR=$(1+ |\Gamma_{load}|) / (1- |\Gamma_{load}|)$ = 3

Smith Chart: Transmission Line Matching, contd.

- Rotate $\lambda/4$ clockwise on circle to Z_n=3 (toward generator by 0.25 λ)
- Impedance seen at that point is Z_n=3, or Z=450
- This is the impedance looking down a $\lambda/4$ length transmission line with Z_0=150 and a load of 50 ohms at the far end
- Here, $\Gamma = (Z-Z_0) / (Z+Z_0)$=0.5 $\angle 0°$
  -Use compass to read Γ on vol. refl. coeff. scale
  -Read angle of Γ off perimeter of Smith chart

Ending point Z_n=3=450/150

Starting point Z_n=50/150=0.33

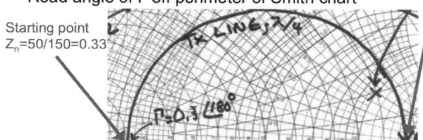

Smith Series R, Parallel G

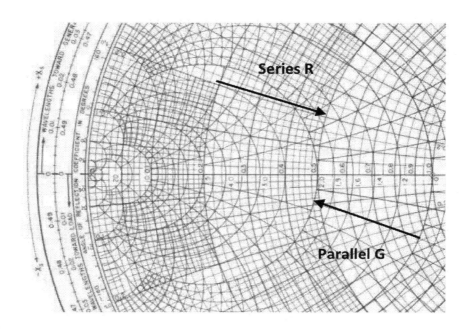

88

Q on Smith Chart

- Q indicates bandwidth
- Lies on arcs in Smith chart
- $Q=|Xs/Rs|$
  $= |Bp/Gp|$
  $= |Rp/Xp|$
- $Q \approx F_0/B_{3dB}$

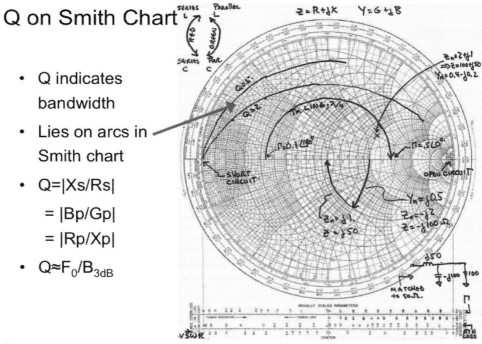

89

Example: Smith Chart Single-Stub Matching 16.7 Ohms

- Starting point at 16.7 ohm load: $Z_n=16.7/50=0.33$ (normalized)
- Rotate 0.083 λ clockwise of series 50-ohm transmission line
- Follow green admittance to center $Z_n=1$, this admittance arc traverses from $B_n=-j1.15$ to $B_n=0$ adding B_n of $j1.15$ corresponding to a capacitive parallel $Z_n=-j/1.15$ or open stub of 0.39-0.25=0.14 λ

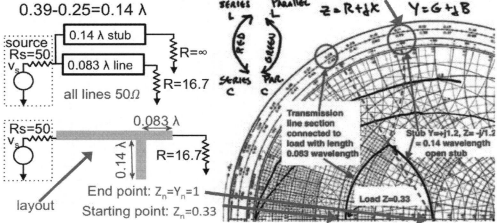

Smith Chart Double-Stub Matching

- Previous example illustrated single-stub impedance matching using the Smith chart
- Double-stub impedance match uses two stubs along the length of a transmission line
- It uses the same basic principles as illustrated in the single-stub example
- However, the procedure is considerable more complicated
- CAD tools are most useful for double-stub impedance match
- General configuration is illustrated below

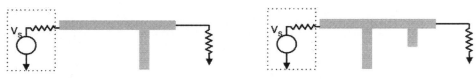

Single-stub impedance match   Double-stub impedance match

Note: stubs may be open or shorted to ground

3D Electromagnetic Simulation

3D Electromagnetic Simulation

- 3D Electromagnetic Simulation
    - o Enter 3D physical structures
    - o Enter material properties
    - o Enter excitation ports
    - o Simulator solves Maxwell's equations
    - o Outputs
        - – S-parameters
        - – Antenna patterns

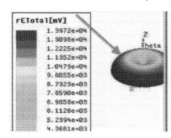

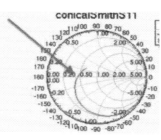

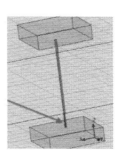

3D Electromagnetic Simulation Waveports

- 3D electromagnetic simulation of coaxial lines
  - Waveports provide input and output ports
  - Added to faces of structures such as end of waveguide
  - After simulation, s-parameters can be plotted

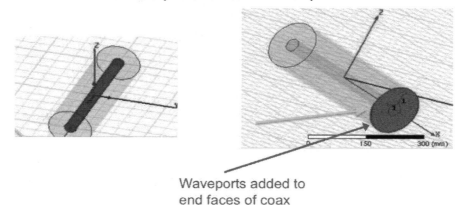

Waveports added to
end faces of coax

3D Electromagnetic Simulation Waveports

- 3D electromagnetic simulation of waveguide
  - Waveports provide input and output ports
  - Added to faces of structures such as end of waveguide
  - After simulation, s-parameters can be plotted

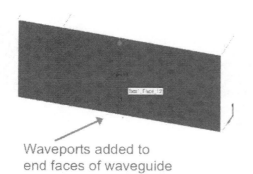

Waveports added to
end faces of waveguide

Rectangular Waveguide

- Rectangular waveguide
- Frequencies below cutoff cannot propagate
  - Cutoff = $1.5 \times 10^8 / a = c/(2a)$
    - where a= longer dimension in meters
    - Smaller dimension is usually approximately a/2
  - WR42: 10.7x4.3mm; 14GHz cutoff
- TE mode, transverse electric field across short dimension
- Only TE01 mode propagates from cutoff to nearly twice cutoff, and the recommended upper operating frequency is less than this

3D Electromagnetic Simulation of Antennas

- 3D Electromagnetic Simulation of antennas
  - Antenna structures can be entered
  - Antenna must have a signal input port
  - After simulation, s-parameters and antenna radiation patterns can be plotted

Antenna Structure

Input Impedance

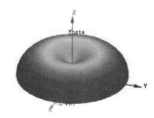

Radiation Pattern

4 STATIC ELECTRIC AND MAGNETIC FIELDS

The lecture notes in this chapter review fundamentals of static electric and magnetic fields
.

Electrostatics and Magnetostatics

Statics: Electrostatics and Magnetostatics

- Electrostatics is the study of constant electric fields
- Magnetostatics is the study of constant magnetic fields
- In statics, the electric and magnetic fields do not change with time
- Therefore, Maxwell's equations have more simple forms for statics
- We will first develop the theory of electrostatics and magnetostatics
- Later, we will develop the full theory of electromagnetics that includes time-varying fields and moving waves, using the full set of Maxwell's equations
- To see where we are eventually going, and to provide insight to what is missing in statics, we will first take a look at the full set of Maxwell's equations for time-varying fields
- For statics, we will remove the time varying parts

Maxwell's Equations

Point Form (Differential form)   Integral Form

$$\nabla \times \mathbf{E}(\overline{x},t) = -\frac{\partial \mathbf{B}(\overline{x},t)}{\partial t} \qquad \text{Faraday's Law} \qquad \oint_L \mathbf{E}(\overline{x},t) \cdot d\mathbf{L} = -\int_S \frac{\partial \mathbf{B}(\overline{x},t)}{\partial t} \cdot d\mathbf{S}$$

assuming $\mathbf{J}_m(\overline{x},t) = 0$

$$\nabla \times \mathbf{H}(\overline{x},t) = \mathbf{J}_e(\overline{x},t) + \frac{\partial \mathbf{D}(\overline{x},t)}{\partial t} \quad \begin{array}{l}\text{Ampere}\\ \text{Circuital Law}\end{array} \quad \oint_L \mathbf{H}(\overline{x},t) \cdot d\mathbf{L} = \int_S \left(\mathbf{J}_e(\overline{x},t) + \frac{\partial \mathbf{D}(\overline{x},t)}{\partial t} \right) \cdot d\mathbf{S}$$

$$\nabla \cdot \mathbf{D}(\overline{x},t) = \rho_e(\overline{x},t) \qquad \text{Gauss' Law} \qquad Q = \int_V \rho_e(\overline{x},t)\,dv = \oint_S \mathbf{D}(\overline{x},t) \cdot d\mathbf{S}$$

$$\nabla \cdot \mathbf{B}(\overline{x},t) = 0 \qquad \text{Gauss' Magnetism Law} \qquad 0 = \oint_S \mathbf{B}(\overline{x},t) \cdot d\mathbf{S}$$

assuming $\rho_m(\overline{x},t) = 0$

$$\mathbf{D}(\overline{x},t) = \varepsilon \mathbf{E}(\overline{x},t) \text{ if } \varepsilon(t) = \delta(t)\varepsilon$$

$$\mathbf{B}(\overline{x},t) = \mu \mathbf{H}(\overline{x},t) \text{ if } \mu(t) = \delta(t)\mu$$

Time varying parts of the equations $\mathbf{J}_e = \sigma \mathbf{E}$

and $\overline{x} = [x\ y\ z]^T$

otherwise

$$\mathbf{D}(\overline{x},t) = \varepsilon(t) * \mathbf{E}(\overline{x},t)$$

$$\mathbf{B}(\overline{x},t) = \mu(t) * \mathbf{H}(\overline{x},t)$$

Maxwell's Equations for Statics (d/dt=0)

- In statics, fields do not change with time, so all d/dt terms =0
- And Maxwell's equations for statics become:

Point Form (Differential form)   Integral Form

$$\nabla \times \mathbf{E}(\overline{x}) = 0 \qquad \text{Kirchoff's Law} \qquad 0 = \oint_L \mathbf{E}(\overline{x}) \cdot d\mathbf{L}$$

$$\nabla \times \mathbf{H}(\overline{x}) = \mathbf{J}_e \qquad \text{Ampere Circuital Law} \quad \oint_L \mathbf{H}(\overline{x}) \cdot d\mathbf{L} = \int_S \mathbf{J}_e(\overline{x}) \cdot d\mathbf{S} = I$$

$$\nabla \cdot \mathbf{D}(\overline{x}) = \rho_e(\overline{x}) \quad \text{Gauss' Law} \quad Q = \int_V \nabla \cdot \mathbf{D}(\overline{x})\,dv = \int_V \rho_e(\overline{x})\,dv = \oint_S \mathbf{D}(\overline{x}) \cdot d\mathbf{S}$$

$$\nabla \cdot \mathbf{B}(\overline{x}) = 0 \qquad \text{Gauss' Magnetism Law} \qquad 0 = \oint_S \mathbf{B}(\overline{x}) \cdot d\mathbf{S}$$

$$\mathbf{D}(\overline{x}) = \varepsilon \mathbf{E}(\overline{x}) \quad \text{Constitutive Equation} \qquad \mathbf{J}_e = \sigma \mathbf{E}$$

$$\mathbf{B}(\overline{x}) = \mu \mathbf{H}(\overline{x}) \quad \text{Constitutive Equation} \qquad \text{and } \overline{x} = [x\ y\ z]^T$$

- Also, there is no phasor form since the Fourier transform of a function that does not change with time only contintains a single frequency component: ω=0, the dc component
- Recall that for constant C, $\mathcal{F}\{$ C $\}=\delta(\omega)$

Shorthand Form: Maxwell's Equations for Statics

- Most commonly, the dependence on spatial coordinates is dropped, and the equations are written as:

Point Form (Differential form)   Integral Form

$$\nabla \times \mathbf{E} = 0 \qquad \text{Kirchoff's Law} \qquad 0 = \oint_L \mathbf{E} \cdot d\mathbf{L}$$

$$\nabla \times \mathbf{H} = \mathbf{J}_e \qquad \text{Ampere Circuital Law} \qquad \oint_L \mathbf{H} \cdot d\mathbf{L} = \int_S \mathbf{J}_e \cdot d\mathbf{S} = I$$

$$\nabla \cdot \mathbf{D} = \rho_e \qquad \text{Gauss' Law} \qquad Q = \int_V \nabla \cdot \mathbf{D}\, dv = \int_V \rho_e\, dv = \oint_S \mathbf{D} \cdot d\mathbf{S}$$

$$\nabla \cdot \mathbf{B} = 0 \qquad \text{Gauss' Magnetism Law} \qquad 0 = \oint_S \mathbf{B} \cdot d\mathbf{S}$$

$\mathbf{D} = \varepsilon \mathbf{E}$ Constitutive Equation $\mathbf{J}_e = \sigma \mathbf{E}$

$\mathbf{B} = \mu \mathbf{H}$ Constitutive Equation

- Again, these are not phasor form since there is no time dependence

Review of Nomenclature

- There is some variation in electromagnetic nomenclature
- For our purposes we will define:
  - o **E** is "electric field intensity" or "E-field" in V/m
  - o **D** is "electric flux density" or "D-field" in C/m²
  - o **H** is "magnetic field intensity" or "H-field" in A/m
  - o **B** is "magnetic flux density" or "B-field" in tesla T or Wb/m²
  - o V is "electric potential" between 2 points in volts
  - o **A** is vector magnetic potential in Wb/m
  - o Beware in electromagnetics literature, because you will see other terms such as "electric field strength," "magnetic field strength," "magnetizing field," "electric field," "magnetic field," and so forth ...so, be careful reading any article/book to find out what terms any author uses, and how they are defined
- Units are often the best way to clear up any confusion
- Safest usage may be "E-field," "B-field," etc.

Vector Operators

Review of Vector Operators

- First, coordinate system must be right-handed
- For now, consider Cartesian coordinates shown
- The dot product (or scalar product) is

$$\mathbf{A} = \begin{bmatrix} A_x \\ A_y \\ A_z \end{bmatrix} = \begin{bmatrix} A_x & A_y & A_z \end{bmatrix}^T \; and \; \mathbf{B} = \begin{bmatrix} B_x & B_y & B_z \end{bmatrix}^T$$

$$\mathbf{A} \cdot \mathbf{B} = \mathbf{A}^T \mathbf{B} = |\mathbf{A}||\mathbf{B}|\cos(\theta) = A_x B_x + A_y B_y + A_z B_z$$

θ is the smaller angle between A and B

NOTE: $\mathbf{A} \cdot \mathbf{A} = A_x^2 + A_y^2 + A_z^2 = |\mathbf{A}|^2$ = length of vector squared

- Vector sum and difference:

$$\mathbf{A} + \mathbf{B} = \begin{bmatrix} A_x + B_x \\ A_y + B_y \\ A_z + B_z \end{bmatrix} \qquad \mathbf{A} - \mathbf{B} = \begin{bmatrix} A_x - B_x \\ A_y - B_y \\ A_z - B_z \end{bmatrix}$$

- Context of vector matters, vectors don't have to "start at origin"
  For example: velocity vector "travels with the particle"

Review of Vector Operators

- Cross product (or vector product)

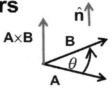

$$\mathbf{A} \times \mathbf{B} = \begin{vmatrix} \hat{\mathbf{x}} & \hat{\mathbf{y}} & \hat{\mathbf{z}} \\ A_x & A_y & A_z \\ B_x & B_y & B_z \end{vmatrix} = |\mathbf{A}||\mathbf{B}|\sin(\theta)\hat{\mathbf{n}}$$

Curl fingers of right hand from **A** to **B** to find direction of A×B

$$= \left(A_y B_z - A_z B_y\right)\hat{\mathbf{x}} - \left(A_x B_z - A_z B_x\right)\hat{\mathbf{y}} + \left(A_x B_y - A_y B_x\right)\hat{\mathbf{z}}$$

and direction of positive angles

where $\hat{n}$ is the unit vector purpendicular to the plane containing **A** and **B** and $\hat{\mathbf{x}}$, $\hat{\mathbf{y}}$, and $\hat{\mathbf{z}}$ are the unit vectors in the direction of the Cartesian axes

- Some properties of dot and cross product:

$$\mathbf{A} \cdot \mathbf{B} = \mathbf{B} \cdot \mathbf{A} \qquad \mathbf{A} \cdot (\mathbf{B} + \mathbf{C}) = \mathbf{A} \cdot \mathbf{B} + \mathbf{A} \cdot \mathbf{C} \qquad \mathbf{A} \cdot \mathbf{A} = |\mathbf{A}|^2$$

$$\mathbf{A} \times \mathbf{B} = -\mathbf{B} \times \mathbf{A} \qquad \mathbf{A} \times (\mathbf{B} + \mathbf{C}) = \mathbf{A} \times \mathbf{B} + \mathbf{A} \times \mathbf{C} \qquad \mathbf{A} \times \mathbf{A} = 0$$

Gradient

- Gradient indicates the direction and magnitude of the change in a scalar field
- The gradient is a vector field calculated from a scalar field
- For a scalar field V(x,y,z,t), the gradient of the field is

define del operator: $\quad \nabla = \hat{\mathbf{x}}\dfrac{\partial}{\partial x} + \hat{\mathbf{y}}\dfrac{\partial}{\partial y} + \hat{\mathbf{z}}\dfrac{\partial}{\partial z} + = \begin{bmatrix} \dfrac{\partial}{\partial x} & \dfrac{\partial}{\partial y} & \dfrac{\partial}{\partial z} \end{bmatrix}^T$

define gradient: $\nabla V\left(\bar{x},t\right) = \hat{\mathbf{x}}\dfrac{\partial}{\partial x}V\left(\bar{x},t\right) + \hat{\mathbf{y}}\dfrac{\partial}{\partial y}V\left(\bar{x},t\right) + \hat{\mathbf{z}}\dfrac{\partial}{\partial z}V\left(\bar{x},t\right)$

$$= \begin{bmatrix} \dfrac{\partial}{\partial x}V\left(\bar{x},t\right) & \dfrac{\partial}{\partial y}V\left(\bar{x},t\right) & \dfrac{\partial}{\partial z}V\left(\bar{x},t\right) \end{bmatrix}^T \quad \text{in Cartesian Coord.}$$

- One of the main uses of gradient is to calculate electric field intensity **E** from electric potential V

$$\nabla V = -\mathbf{E} \qquad \text{For static E-field only}$$

54

Divergence

- Divergence indicates total outward flow of a vector field for an infinitesimal volume
- Divergence of a vector field **D**(x,y,z,t) is:

$$\text{define divergence:} \quad \nabla \cdot \mathbf{D} = \begin{bmatrix} \dfrac{\partial}{\partial x} & \dfrac{\partial}{\partial y} & \dfrac{\partial}{\partial z} \end{bmatrix} \cdot \begin{bmatrix} D_x(\bar{x},t) \\ D_y(\bar{x},t) \\ D_z(\bar{x},t) \end{bmatrix}$$

$$= \frac{\partial}{\partial x} D_x(\bar{x},t) + \frac{\partial}{\partial y} D_y(\bar{x},t) + \frac{\partial}{\partial z} D_z(\bar{x},t) \text{ in Cartesian Coord.}$$

- The field shown below has zero divergence everywhere except at the center

  Field with zero divergence everywhere but at center point. Divergence is zero outside.
  Caution: not every region where field lines "move apart" has non-zero divergence

$$\nabla \cdot \mathbf{D} = \rho_e$$

Curl

- Curl indicates rotation about a point in a vector field
- For a vector field **B**(x,y,z,t), the curl of the field is:

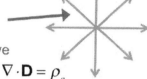

Field without curl

Field with nonzero curl inside conductor. But curl is zero outside. **Caution**: not every curved field has non-zero curl

$$\text{define curl:}$$

$$\nabla \times \mathbf{B} = \begin{vmatrix} \hat{\mathbf{x}} & \hat{\mathbf{y}} & \hat{\mathbf{z}} \\ \dfrac{\partial}{\partial x} & \dfrac{\partial}{\partial y} & \dfrac{\partial}{\partial z} \\ B_x(\bar{x},t) & B_y(\bar{x},t) & B_z(\bar{x},t) \end{vmatrix}$$

$$= \left(\frac{\partial}{\partial y} B_z(\bar{x},t) - \frac{\partial}{\partial z} B_y(\bar{x},t) \right)\hat{\mathbf{x}} - \left(\frac{\partial}{\partial x} B_z(\bar{x},t) - \frac{\partial}{\partial z} B_x(\bar{x},t) \right)\hat{\mathbf{y}}$$

$$+ \left(\frac{\partial}{\partial x} B_y(\bar{x},t) - \frac{\partial}{\partial y} B_x(\bar{x},t) \right)\hat{\mathbf{z}}$$

and $\hat{\mathbf{x}}$, $\hat{\mathbf{y}}$, and $\hat{\mathbf{z}}$ are the unit vectors in the direction of the Cartesian axes

Laplacian, Poisson's Eq., and Laplace's Eq.

- The Laplacian of a scalar field serves a role similar to second derivative

- It is often used with scalar potential (voltage)

$$\text{define Laplacian: } \nabla^2 V = \nabla \cdot \nabla V = \begin{bmatrix} \dfrac{\partial}{\partial x} & \dfrac{\partial}{\partial y} & \dfrac{\partial}{\partial z} \end{bmatrix} \cdot \begin{bmatrix} \dfrac{\partial}{\partial x} \\ \dfrac{\partial}{\partial y} \\ \dfrac{\partial}{\partial z} \end{bmatrix} V(\bar{x}, t)$$

$$= \left(\frac{\partial^2}{\partial x^2} + \frac{\partial^2}{\partial y^2} + \frac{\partial^2}{\partial z^2} \right) V(\bar{x}, t) = -\nabla \cdot \mathbf{E} = -\left(\frac{\partial E_x}{\partial x} + \frac{\partial E_y}{\partial y} + \frac{\partial E_z}{\partial z} \right)$$

- The Laplacian applied to the scalar potential results in Poisson's equation and Laplace's equation:

$$\nabla^2 V = \nabla \cdot \nabla V = -\nabla \cdot \mathbf{E} = -\nabla \cdot \mathbf{D}/\varepsilon = -\rho_e/\varepsilon$$

$\Rightarrow$ Poisson's Eq.: $\nabla^2 V = -\rho_e/\varepsilon$ For static E-field only

$\Rightarrow$ Laplace's Eq.: $\nabla^2 V = 0$ for charge-free regions

Laplacian of a Vector Field

- The Laplacian of a vector field differs from the Laplacian of a scalar field, but still is similar to a second derivative

Laplacian of a vector follows from vector identity:

$$\nabla^2 \mathbf{E} = \nabla(\nabla \cdot \mathbf{E}) - \nabla \times \nabla \times \mathbf{E}$$

$$= \nabla \left(\frac{\partial E_x}{\partial x} + \frac{\partial E_y}{\partial y} + \frac{\partial E_z}{\partial z} \right) - \begin{vmatrix} \hat{\mathbf{x}} & \hat{\mathbf{y}} & \hat{\mathbf{z}} \\ \dfrac{\partial}{\partial x} & \dfrac{\partial}{\partial y} & \dfrac{\partial}{\partial z} \\ \left(\dfrac{\partial}{\partial y} E_z - \dfrac{\partial}{\partial z} E_y \right) & \left(\dfrac{\partial}{\partial z} E_x - \dfrac{\partial}{\partial x} E_z \right) & \left(\dfrac{\partial}{\partial x} E_y - \dfrac{\partial}{\partial y} E_x \right) \end{vmatrix}$$

$$= \begin{bmatrix} \nabla^2 E_x & \nabla^2 E_y & \nabla^2 E_z \end{bmatrix}^T$$

- The Laplacian of a vector field will later be useful in finding the wave equation

Einstein Notation (slightly modified)

- Many vector operations can be described in Einstein notation
- Einstein notation: rule is that repeated indices are summed
- Dot product

$$\mathbf{A} = \begin{bmatrix} A_x & A_y & A_z \end{bmatrix}^T = \begin{bmatrix} a_1 & a_2 & a_3 \end{bmatrix}^T \ and \ \mathbf{B} = \begin{bmatrix} b_1 & b_2 & b_3 \end{bmatrix}^T$$

$$\mathbf{A} \cdot \mathbf{B} = \text{in Einstein notation} = a_i b_i = \sum_{i=1}^{3} a_i b_i = a_1 b_1 + a_2 b_2 + a_3 b_3$$

- Cross product in Einstein notation:

$$define: \begin{bmatrix} \hat{\mathbf{x}} & \hat{\mathbf{y}} & \hat{\mathbf{z}} \end{bmatrix}^T = \begin{bmatrix} \hat{\mathbf{e}}_1 & \hat{\mathbf{e}}_2 & \hat{\mathbf{e}}_3 \end{bmatrix}^T$$

$$\mathbf{A} \times \mathbf{B} = \text{in Einstein notation} = \epsilon_{ijk} \hat{\mathbf{e}}_i a_j b_k = \sum_{i=1}^{3} \sum_{j=1}^{3} \sum_{k=1}^{3} \epsilon_{ijk} \hat{\mathbf{e}}_i a_j b_k$$

$$\text{where Levi-Civita symbol } \epsilon_{ijk} = \begin{cases} 1 & if & (i,j,k) = (1,2,3), (3,1,2), or (2,3,1) \\ -1 & if & (i,j,k) = (3,2,1), (1,3,2), or (2,1,3) \\ 0 & & otherwise \end{cases}$$

Divergence Theorem

- Divergence Theorem.
- For any vector field $\mathbf{A}(x,y,z,t)$, the volume integral of the divergence equals the closed-surface integral of the dot product $\mathbf{A} \cdot d\mathbf{L}$ over the surface with the outward differential surface normal $d\mathbf{S}$

$$\text{Divergence Theorem:} \int_V \nabla \cdot \mathbf{A}(\overline{x}, t) \, dv = \oint_S \mathbf{A}(\overline{x}, t) \cdot d\mathbf{S}$$

$$\text{Point form of Gauss' Law:} \quad \nabla \cdot \mathbf{D}(\overline{x}, t) = \rho_e(\overline{x}, t)$$

$$\text{substituting yields:} \quad Q = \int_V \rho_e(\overline{x}, t) \, dv = \oint_S \mathbf{D}(\overline{x}, t) \cdot d\mathbf{S}$$

- The fields and regions are illustrated below
- Surface normal $d\mathbf{S}$ is outward facing

Surface element dS 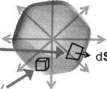 dS

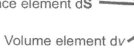

Volume element dv

Stokes' Theorem

- Stokes' Theorem. The open surface integral of the curl of any vector field $\mathbf{A}(x,y,z,t)$ equals the closed line integral of the dot product $\mathbf{A} \cdot d\mathbf{L}$ along the perimeter L of the open surface

$$\int_S \nabla \times \mathbf{A}(\overline{x},t) \cdot d\mathbf{S} = \oint_L \mathbf{A}(\overline{x},t) \cdot d\mathbf{L}$$

Ampere Circuital Law $\quad \nabla \times \mathbf{H}(\overline{x},t) = \mathbf{J}_e(\overline{x},t) + \dfrac{\partial \mathbf{D}(\overline{x},t)}{\partial t}$

$$\Rightarrow I = \int_S \nabla \times \mathbf{H}(\overline{x},t) \cdot d\mathbf{S} = \int_S \left(\mathbf{J}_e(\overline{x},t) + \frac{\partial \mathbf{D}(\overline{x},t)}{\partial t} \right) \cdot d\mathbf{S} = \oint_L \mathbf{H}(\overline{x},t) \cdot d\mathbf{L}$$

- The fields and regions are illustrated below
- Surface normal d**S**
- Positive d**S** follows right hand rule of d**L**
- L is the closed path around the perimeter

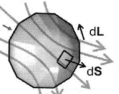

Important Identities and Properties

- The curl of the gradient of a scalar field is zero:
$$\nabla \times (\nabla V) = 0$$
- And, it is also true that if $\nabla \times \mathbf{A} = 0$, there exists some scalar field B such that $\nabla B = \mathbf{A}$
- A vector field is a conservative field if the line integral between two points is independent of the path L:
$$\int_{L1} \mathbf{A} \cdot d\mathbf{L} = \int_{L2} \mathbf{A} \cdot d\mathbf{L} \quad \text{for all paths L1 and L2, or equivalently } \oint_L \mathbf{A} \cdot d\mathbf{L} = 0$$
- Using Stokes theorem, all fields with zero curl are conservative:
$$\int_S \nabla \times \mathbf{A} \cdot d\mathbf{S} = \oint_L \mathbf{A} \cdot d\mathbf{L} = 0 \quad \text{if } \nabla \times \mathbf{A} = 0$$
- Another useful vector identity is:
$$\nabla \cdot (\nabla \times \mathbf{A}) = 0$$
- And so, any vector field with $\nabla \cdot \mathbf{B} = 0$ can be written as the curl of another vector field $\mathbf{A}$ where $\nabla \times \mathbf{A} = \mathbf{B}$
- Since $\mathbf{E} = -\nabla V$, all *static* E-fields are conservative

Electrostatics

Coulomb's Law

- Coulomb's Law describes the electrostatic force vector $\mathbf{F}_1$ N on particle 1 of charge q_1 C, and force $\mathbf{F}_2$ N on particle 2 of charge q_2, with vector length $\mathbf{R}_{12}$ separation directed from q_1 to q_2

$$\mathbf{F}_2 = \frac{q_1 q_2}{4\pi\varepsilon} \frac{\mathbf{R}_{12}}{|\mathbf{R}_{12}|^3} = \frac{q_1 q_2}{4\pi\varepsilon_r \varepsilon_0} \frac{\mathbf{R}_{12}}{|\mathbf{R}_{12}|^3} = q_2 \mathbf{E}_1 = -\mathbf{F}_1$$

$$\mathbf{E}_1 = \frac{q_1}{4\pi\varepsilon} \frac{\mathbf{R}_{12}}{|\mathbf{R}_{12}|}$$

$where: \varepsilon_0 = 8.85$ pF/m

This is the basic definition of E-field

- Example: find the forces on two positive charges of 1 C separated by 1 m in vacuum

$$\mathbf{F}_2 = -\mathbf{F}_1 = \frac{q_1 q_2}{4\pi\varepsilon_r \varepsilon_0} \frac{\mathbf{R}_{12}}{|\mathbf{R}_{12}|^3}$$

$$= \frac{1}{4\pi(8.85\times10^{-12})}\hat{\mathbf{y}} = 9.0\times10^9 \hat{\mathbf{y}} \ N$$

E-field of a Point Charge

- As noted on the previous slide, the definition for the electric field intensity **E** is taken from the Coulomb force
- The electric field intensity **E** is defined as the force experienced by a point charge of 1 C
- Recall that the force on charge q_2 is

$$\mathbf{F}_2 = \frac{q_1 q_2}{4\pi\varepsilon} \frac{\mathbf{R}_{12}}{|\mathbf{R}_{12}|^3}$$ and note that $$|\mathbf{F}_2| = \frac{q_1 q_2}{4\pi\varepsilon |\mathbf{R}_{12}|^2}$$ is an inverse-square law

- So, setting q_2 =1 C, results in the formula for the field **E** produced by q_1

$$\mathbf{E} = \frac{q_1}{4\pi\varepsilon} \frac{\mathbf{R}_{12}}{|\mathbf{R}_{12}|^3}$$ Note that |**E**| follows an inverse-square law

- The E-field is plotted as shown
- Note that the density of the lines indicates the strength of the field

Gauss' Law

- Gauss' Law integral form relates the electrostatic charge Q that is contained in a volume V to the closed surface integral of the dot product **D·dS** over the surface with the outward differential surface normal d**S**

Point Form (Differential form)   Integral Form

$$\nabla \cdot \mathbf{D} = \rho_e$$ Gauss' Law $$Q = \int_V \nabla \cdot \mathbf{D}\, dv = \int_V \rho_e(\bar{x})\, dv = \oint_S \mathbf{D} \cdot d\mathbf{S}$$

$$\mathbf{D} = \varepsilon_r \varepsilon_0 \mathbf{E} = \varepsilon \mathbf{E}$$ Constitutive Eq., where $\varepsilon_0 = 8.85$ pF/m and $\bar{x} = [x\ y\ z]^T$

- Example: a sphere of radius 1 m has uniform outward electric field E=1 V/m in vacuum, find the interior <u>uniform</u> charge density

$$Q = \oint_S \mathbf{D} \cdot d\mathbf{S} = \oint_S \varepsilon_0 \mathbf{E} \cdot d\mathbf{S} = \oint_S \varepsilon_0\, ds = 4\pi r^2 \varepsilon_0 = 4\pi(8.85\times10^{-12}) = 111 \text{ pC}$$

$$Q = 111 \text{ pC} = \int_V \rho_e(\bar{x})\, dv = \frac{4\pi r^3}{3}\rho_e$$

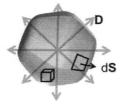

$$so: \rho_e = \frac{3}{4\pi r^3}Q = \frac{3}{4\pi}111\times10^{-12} = 26.5 \text{ pC/m}^3$$

Continuity of Current

- Gauss' Law can be used to determine necessary relations for the continuity of current and conservation of charge
- Consider the change in charge I=dQ/dt contained in a volume V that is caused by vector current density $\mathbf{J_e}$ crossing the surface of the volume, as illustrated below

$$I_{out} = -\frac{dQ}{dt} = -\frac{d}{dt}\int_V \rho_e \, dv = -\int_V \frac{\partial}{\partial t}\rho_e \, dv = \oint_S \mathbf{J_e} \cdot d\mathbf{S}$$

- Recall the divergence theorem

$$\text{Divergence Theorem:} \quad \int_V \nabla \cdot \mathbf{A} \, dv = \oint_S \mathbf{A} \cdot d\mathbf{S}$$

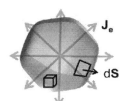

- Then $\quad I_{out} = -\int_V \frac{\partial}{\partial t}\rho_e \, dv = \oint_S \mathbf{J_e} \cdot d\mathbf{S} = \int_V \nabla \cdot \mathbf{J} \, dv$

- Comparing sides, the point form of continuity is

$$\nabla \cdot \mathbf{J} = -\frac{\partial}{\partial t}\rho_e$$

Static Electric Potential V

- A 1 V electric potential between 2 points in space means that 1 J of external energy is required to move 1 C of charge from the point at lower potential to the point with higher potential
- In the opposite direction, 1J of energy is delivered externally
- The **static** electric potential V between point A and point B in an electric field is found from the line integral of the dot product $\mathbf{E} \cdot d\mathbf{L}$ from point A to point B

$$V_{AB} = V_A - V_B = -\int_L \mathbf{E} \cdot d\mathbf{L} = -\int_B^A \mathbf{E} \cdot d\mathbf{L} = \int_A^B \mathbf{E} \cdot d\mathbf{L} \quad \text{and} \quad \oint_L \mathbf{E} \cdot d\mathbf{L} = 0$$

- As noted on the right above, any closed path returns to the original point and returns to the original electric potential V
- Also, V does not depend on the particular path taken
- Finally, the gradient of V yields the electric field

$$\mathbf{E} = -\nabla V$$

For static E-field only

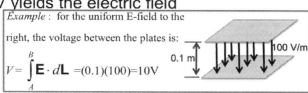

Example : for the uniform E-field to the right, the voltage between the plates is:

$$V = \int_A^B \mathbf{E} \cdot d\mathbf{L} = (0.1)(100) = 10\text{V}$$

0.1 m

100 V/m

Plotting V and E

- Voltage V_{AB} is defined as a potential difference, so some arbitrary point is chosen as "V=0" as the reference point B for referencing all voltages (potential energies)

- Often, a point at infinity is a useful "V=0" reference point

$$V = V_{A\infty} = V_A - V_\infty = -\int_\infty^A \mathbf{E} \cdot d\mathbf{L} = \int_A^\infty \mathbf{E} \cdot d\mathbf{L} \text{ work moving 1C from } \infty \text{ to A}$$

- For example, consider the potential V referenced at infinity for a point charge q centered at the origin

$$V = V_{A\infty} = V_A - V_\infty = -\int_\infty^A \frac{q}{4\pi\varepsilon} \frac{\mathbf{R}}{|\mathbf{R}|^3} \cdot d\mathbf{L}$$

$$= -\int_\infty^r \frac{q}{4\pi\varepsilon} \frac{r\hat{\mathbf{r}}}{r^3} \cdot dr\,\hat{\mathbf{r}} = -\int_\infty^A \frac{q}{4\pi\varepsilon r^2} dr$$

$$= \frac{q}{4\pi\varepsilon r_A} - \frac{q}{4\pi\varepsilon\infty} = \frac{q}{4\pi\varepsilon r_A}$$

Equipotential surfaces

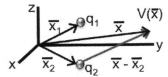

$V_1 = V_{r=1}$
$V_1/2$
$V_1/3$

Static Electric Potential Due To Collection of Charges

- An electric potential V referenced at infinity for a collection of N point charges is

$$V(\bar{x}) = \sum_{n=1}^N \frac{q_n}{4\pi\varepsilon_0 |\bar{x} - \bar{x}_n|} \quad where \quad V(\infty) = 0, and \ \bar{x} = [x\ y\ z]^T$$

and $\bar{x}_n$ is the vector coordinates of charge q_n

- For a volume of charge density this becomes an integral

$$V(\bar{x}) = \int_V \frac{\rho_e(\bar{\alpha})}{4\pi\varepsilon_0 |\bar{x} - \bar{\alpha}|} dv \quad where \quad V(\infty) = 0, and \ \bar{x} = [x\ y\ z]^T$$

$\bar{\alpha}$ is coordinate of charge density

- More integrals exist in the literature for line and surface charge

- Finally, the gradient of V yields the E-field

$$\mathbf{E} = -\nabla V \quad \text{For static E-field only}$$

Poisson's Equation and Laplace's Equation

- The Laplacian applied to the E-field results in Poisson's equation and Laplace's equation:

$$\nabla^2 V = \nabla \cdot \nabla V = -\nabla \cdot \mathbf{E} = -\nabla \cdot \mathbf{D}/\varepsilon = -\rho_e/\varepsilon$$

$$\Rightarrow \text{Poisson's Eq.:} \quad \nabla^2 V = -\rho_e/\varepsilon \qquad \text{For static E-field only}$$

- In charge-free regions, Poisson's equation becomes Laplace's equation:

$$\text{Poisson's Eq.:} \quad \nabla^2 V = -\rho_e/\varepsilon \qquad \text{For static E-field only}$$

$$\Rightarrow \text{Laplace's Eq.:} \quad \nabla^2 V = 0 \qquad \text{for charge-free regions}$$

Capacitance

- Capacitance relates the amount of charge stored on a device with an electric potential of 1 V applied, thus:

$$C = \frac{Q}{V}$$

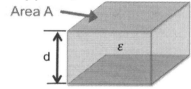

Area A

and for parallel plates $C = \dfrac{\varepsilon A}{d} = \dfrac{\varepsilon_r \varepsilon_0 A}{d}$

where A is plate area and d is plate separation

- Capacitance is also a measure of energy storage, since the energy used to reach a voltage V is

$$energy = W_E = \int_0^V V\, dQ = \int_0^V V(C\, dV) = \frac{CV^2}{2}$$

- Example: compute the capacitance of a parallel plate capacitor shown below, with a 2x2 mm square plate separated by 1 mm in vacuum

$$C = \frac{\varepsilon A}{d} = \frac{8.85 \times 10^{-12}(0.002)^2}{0.001} = 35.4 \text{ fF}$$

Conduction

- Conductivity σ in S/m is a property of conductors and determines the current density $\mathbf{J_e}$ in a material as a function of the E-field present
- In the presence of an E-field $\mathbf{E}$, the current density is

$$\mathbf{J_e} = \sigma\mathbf{E}$$

- Notice that the resulting current density $\mathbf{J_e}$ is in the same direction as the E-field $\mathbf{E}$ that produces the current

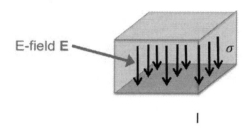

Resistance

- Resistance equals 1/i where i is the amount of current through a device with an electric potential of 1 V applied, thus, for the device below with top and bottom metal plates of area A, separated by distance d, and material of conductivity σ between the plates:

$$R = \frac{V}{I} = \frac{\int_L \mathbf{E}\cdot d\mathbf{L}}{\int_S \mathbf{J_e}\cdot d\mathbf{S}} = \frac{\int_L \mathbf{E}\cdot d\mathbf{L}}{\int_S \sigma\mathbf{E}\cdot d\mathbf{S}}$$

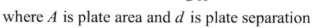

and for parallel plates $R = \dfrac{d}{\sigma A}$

where A is plate area and d is plate separation

- Example: compute the resistance of a 1x1 mm square copper wire of length 100 m and conductivity $\sigma=6\times10^7$ S/m

$$R = \frac{d}{\sigma A} = \frac{100}{(6\times10^7)(0.001)^2} = 1.7 \text{ ohms}$$

Magnetostatics

Biot-Savart Law

- The Biot-Savart Law describes the incremental contribution d**H** to the H-field at distance **R** from differential current element I d**L**

$$dH = \frac{I\,d\mathbf{L} \times \mathbf{R}}{4\pi |\mathbf{R}|^3}$$

- Example: find the H-field in the center of a 2 m radius current loop in the x-y plane, centered on the z-axis in vacuum

$$\mathbf{H} = \int d\mathbf{H} = \int \frac{I\,d\mathbf{L} \times \mathbf{R}}{4\pi |\mathbf{R}|^3} = \int_0^{2\pi} \frac{I\,Rd\varphi(R\hat{\mathbf{z}})}{4\pi R^3}$$

$$= \int_0^{2\pi} \frac{I\,d\varphi\hat{\mathbf{z}}}{4\pi R} = \frac{2\pi I}{4\pi R}\hat{\mathbf{z}} = \frac{I}{2R}\hat{\mathbf{z}} = \frac{1}{4}\hat{\mathbf{z}}\ \text{A/m}$$

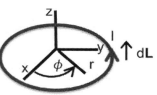

Gauss' Law for Magnetism

- Gauss' Law for magnetism states that the closed surface integral of the dot product **B·dS** over the surface with the outward differential surface normal d**S** is equal to zero:

Point Form (Differential form)　　　　　　　　　　Integral Form

$$\nabla \cdot \mathbf{B} = 0 \qquad \text{Gauss' Law} \qquad 0 = \int_V \rho_m(\bar{x})\,dv = \oint_S \mathbf{B} \cdot d\mathbf{S}$$

$$\mathbf{B} = \mu_r\mu_0\mathbf{H} = \mu\mathbf{H}, \quad \rho_m(\bar{x}) = 0, \quad \mu_0 = 1257 \text{ nH/m} \ \text{ and } \ \bar{x} = [x\ y\ z]^T$$

- Example: show that Gauss's Magnetism Law is satisfied for a wire of radius 1 cm carrying a current current I with an external magnetic field intensity |H|=I/(2πr) outside the wire

choose the cylinder surface S

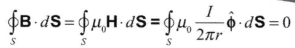

$$\oint_S \mathbf{B} \cdot d\mathbf{S} = \oint_S \mu_0\mathbf{H} \cdot d\mathbf{S} = \oint_S \mu_0 \frac{I}{2\pi r}\hat{\phi} \cdot d\mathbf{S} = 0$$

since $\hat{\phi}$ is orthogonal to all the $d\mathbf{S}$ of the cylinder surface

Ampere's Circuital Law for Magnetostatics

- Ampere's Circuital Law for Magnetostatics (dropping the time-varying terms) states that the open-surface integral of the current density **J**$_e$= $\nabla \times$**H** both equals the current through the surface, and also equals the closed line integral of the dot product **H·dL** along the perimeter L of the open surface

Point Form (Differential form)　　　　　　　　Integral Form

$$\nabla \times \mathbf{H} = \mathbf{J}_e \qquad\qquad \oint_L \mathbf{H} \cdot d\mathbf{L} = \int_S \nabla \times \mathbf{H} \cdot d\mathbf{S} = \int_S \mathbf{J}_e \cdot d\mathbf{S} = I$$

- Example: copper wire of 1 mm radius has an internal electric field 10^{-6} V/m with $\sigma = 6 \times 10^7$ S/m and external H-field of |H|=I/(2πr) outside the wire.

choose the disk surface S

$$I = \int_S \mathbf{J}_e \cdot d\mathbf{S} = \int_S \sigma\mathbf{E} \cdot d\mathbf{S} = \int_S \sigma|\mathbf{E}|\hat{z} \cdot ds\hat{z} = |\mathbf{E}|2\pi r_{wire}^2 \sigma$$

$$= (10^{-6})2\pi(0.001)^2(6 \times 10^7) = 37.7 \text{ A}$$

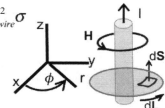

$$\oint_L \mathbf{H} \cdot d\mathbf{L} = \oint_L \frac{I}{2\pi r}\hat{\phi} \cdot dl\hat{\phi} = I \qquad \text{Since |H|=I/2πr}$$
outside the wire

Inductance

- Inductance equals the amount of magnetic flux in a device with an current of 1 A applied, thus:

and for a solenoid with N turns $|\mathbf{B}| \approx \dfrac{\mu_r \mu_0 N I}{d}$

$$L = \frac{\Phi}{I} = \frac{\int_S \mathbf{B} \cdot d\mathbf{S}}{I} \approx \frac{\int_S \frac{\mu_r \mu_0 N I}{d} \hat{\mathbf{z}} \cdot ds\hat{\mathbf{z}}}{I} = \frac{N 2\pi r^2 \mu_r \mu_0}{d}$$

where $\mu_0 = 1257$ nH/m

- The energy stored in an inductor is:

$$energy = W_M = \frac{L I^2}{2}$$

- Example: compute the inductance of a 10-turn solenoid with 1 mm radius and 5 mm length in vacuum

$$L = \frac{N^2 \pi r^2 \mu_r \mu_0}{d} = \frac{(10)^2 \pi (0.001)^2 (1257 \times 10^{-9})}{(0.005)} = 79 \text{ nH}$$

Energy Stored in Static Fields

- Energy is stored in both the electrostatic field and the magnetostatic field.

- The energy-density w_E in J/m^3 and energy W_E in J within a volume V of the electrostatic field is:

$$w_E = \frac{\varepsilon |\mathbf{E}|^2}{2} = \frac{\varepsilon_r \varepsilon_0 |\mathbf{E}|^2}{2} \quad and \quad W_E = \int_V w_e(\overline{x}) dv = \frac{1}{2} \int_V \varepsilon |\mathbf{E}|^2 dv = \frac{1}{2} \int_V \mathbf{D} \cdot \mathbf{E} dv$$

where $\varepsilon_0 = 8.85$ pF/m

- The energy-density w_M in J/m^3 and energy W_M in J within a volume V of the magnetostatic field is:

$$w_M = \frac{\mu |\mathbf{H}|^2}{2} = \frac{\mu_r \mu_0 |\mathbf{H}|^2}{2} \quad and \quad W_M = \int_V w_M(\overline{x}) dv = \frac{1}{2} \int_V \mu |\mathbf{H}|^2 dv = \frac{1}{2} \int_V \mathbf{B} \cdot \mathbf{H} dv$$

where $\mu_0 = 1257$ nH/m

Lorentz Force

- In addition to the electrostatic force of Coulomb's Law, the magnetic flux density produces and additional force on moving charges, even charge moving inside a conductor
- In the absence of electric field, the force **F** on a charge with velocity vector **U** in a magnetic field is

$$\mathbf{F} = q\mathbf{U} \times \mathbf{B}$$

- The total force in combined electric and magnetic fields is called the Lorentz Force, given below:

$$\mathbf{F} = q\left(\mathbf{E} + \mathbf{U} \times \mathbf{B}\right)$$

- Example: find the force on 10^{-3} C charged sphere moving at 100 m/s in a 1 Tesla B-field, in the configuration above

$$\mathbf{F} = q\left(\mathbf{E} + \mathbf{U} \times \mathbf{B}\right) = q\mathbf{U} \times \mathbf{B} = q|\mathbf{U}||\mathbf{B}|\sin(\theta)\hat{\mathbf{z}}$$

$$= (0.001)(100)(1)\hat{\mathbf{z}} = 0.1\hat{\mathbf{z}} \text{ N}$$

Static (dc) Vector Magnetic Potential A

- If we try to define a scalar magnetic potential similar to electric potential $\mathbf{E} = -\nabla V$, there is a problem since $\nabla \times \mathbf{H} = \mathbf{J_e}$, but vector identity $\nabla \times (\nabla X) = 0$ for any scalar field X.
- The solution is to define a vector magnetic potential **A** (Wb/m)

$$\mathbf{B} = \nabla \times \mathbf{A} \quad so \quad \nabla \times \mathbf{H} = \frac{1}{\mu}\nabla \times \left(\nabla \times \mathbf{A}\right) = \mathbf{J_e}$$

- The potential A can be determined from the currents, similar to the way voltage potential is found from the charge distribution

$$\mathbf{A}(\bar{x}) = \oint_L \frac{\mu I \, d\mathbf{L}}{4\pi|\bar{x} - \bar{\alpha}|} \quad \text{compare to voltage} \quad V(\bar{x}) = \int_V \frac{\rho_e(\bar{\alpha})}{4\pi\varepsilon_0|\bar{x} - \bar{\alpha}|}dv$$

$\bar{\alpha}$ is coordinate of $d\mathbf{L}$ and $\mathbf{A}(\infty) = 0$, and $\bar{x} = [x \ y \ z]^T$

- Finally, the curl of **A** yields the B-field

$$\mathbf{B} = \nabla \times \mathbf{A}$$

- Note: for a volume

$$\mathbf{A}(\bar{x}) = \int_V \frac{\mu \mathbf{J} \, dv}{4\pi|\bar{x} - \bar{\alpha}|}$$

Let dA be the integrand

Note: dA Is in direction of dL

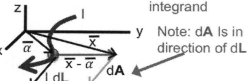

Shorthand Form: Maxwell's Equations for Statics

- Most commonly, the dependence on spatial coordinates is dropped, and the equations are written as:

Point Form (Differential form)   Integral Form

$$\nabla \times \mathbf{E} = 0 \qquad \text{Kirchoff's Law} \qquad 0 = \oint_L \mathbf{E} \cdot d\mathbf{L}$$

$$\nabla \times \mathbf{H} = \mathbf{J_e} \qquad \text{Ampere Circuital Law} \qquad \oint_L \mathbf{H} \cdot d\mathbf{L} = \int_S \mathbf{J_e} \cdot d\mathbf{S} = I$$

$$\nabla \cdot \mathbf{D} = \rho_e \qquad \text{Gauss' Law} \qquad Q = \int_V \nabla \cdot \mathbf{D} \, dv = \int_V \rho_e \, dv = \oint_S \mathbf{D} \cdot d\mathbf{S}$$

$$\nabla \cdot \mathbf{B} = 0 \qquad \text{Gauss' Magnetism Law} \qquad 0 = \oint_S \mathbf{B} \cdot d\mathbf{S}$$

$\mathbf{D} = \varepsilon \mathbf{E}$ Constitutive Equation $\mathbf{J_e} = \sigma \mathbf{E}$

$\mathbf{B} = \mu \mathbf{H}$ Constitutive Equation

- Again, these are not phasor form since there is no time dependence

5 TIME-VARYING FIELDS

The lecture notes in this chapter present analysis of time-varying fields.

From Statics to Time-Varying Fields

- In statics, the electric and magnetic fields did not change with time
- This allowed use of a more simple form of Maxwell equations, dropping d/dt terms
- Next, we will develop the full theory of electromagnetics that includes time-varying fields and moving waves,
- But first, we will quickly review the terms that were discarded in statics
- We will address these time-varying components today

Maxwell's Equations

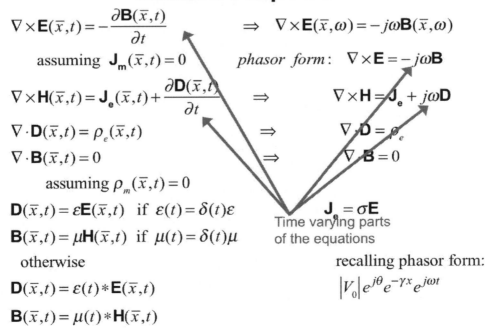

$$\nabla \times \mathbf{E}(\overline{x},t) = -\frac{\partial \mathbf{B}(\overline{x},t)}{\partial t}$$

assuming $\mathbf{J}_m(\overline{x},t) = 0$

$$\Rightarrow \quad \nabla \times \mathbf{E}(\overline{x},\omega) = -j\omega\mathbf{B}(\overline{x},\omega)$$

phasor form: $\nabla \times \mathbf{E} = -j\omega\mathbf{B}$

$$\nabla \times \mathbf{H}(\overline{x},t) = \mathbf{J}_e(\overline{x},t) + \frac{\partial \mathbf{D}(\overline{x},t)}{\partial t}$$

$$\Rightarrow \quad \nabla \times \mathbf{H} = \mathbf{J}_e + j\omega\mathbf{D}$$

$$\nabla \cdot \mathbf{D}(\overline{x},t) = \rho_e(\overline{x},t)$$

$$\Rightarrow \quad \nabla \cdot \mathbf{D} = \rho_e$$

$$\nabla \cdot \mathbf{B}(\overline{x},t) = 0$$

$$\Rightarrow \quad \nabla \cdot \mathbf{B} = 0$$

assuming $\rho_m(\overline{x},t) = 0$

$$\mathbf{D}(\overline{x},t) = \varepsilon\mathbf{E}(\overline{x},t) \quad \text{if} \quad \varepsilon(t) = \delta(t)\varepsilon$$

$$\mathbf{B}(\overline{x},t) = \mu\mathbf{H}(\overline{x},t) \quad \text{if} \quad \mu(t) = \delta(t)\mu$$

$$\mathbf{J}_e = \sigma\mathbf{E}$$

Time varying parts of the equations

otherwise

recalling phasor form:

$$\mathbf{D}(\overline{x},t) = \varepsilon(t)*\mathbf{E}(\overline{x},t)$$

$$|V_0|e^{j\theta}e^{-\gamma x}e^{j\omega t}$$

$$\mathbf{B}(\overline{x},t) = \mu(t)*\mathbf{H}(\overline{x},t)$$

Faraday's Law

This is the first law where we see interaction between electric and magnetic fields

Faraday's Law

- Faraday's law describes the emf (electromotive force) voltage that is induced when magnetic flux Φ in a loop changes

$$\text{emf} = -\frac{d\Phi}{dt} = -\frac{\partial}{\partial t}\int_S \mathbf{B} \cdot d\mathbf{S} = \oint_L \mathbf{E} \cdot d\mathbf{L}$$

$$also = -\int_S \frac{\partial \mathbf{B}}{\partial t} \cdot d\mathbf{S} \quad \text{if surface S doesnt change}$$

Note: no "-" sign

Important: Positive senses of **B**, dS and dL per right-hand rule

- Note that the change in magnetic flux $d\Phi/dt$ can occur through the change in the external magnetic flux, motion of the coil, motion of magnets, and so forth.

- Also, the closed loop line integral of E-field is no longer zero, as for electrostatics. This E-field is created by $d\Phi/dt$ not charge!

- If there are N turns in the loop, emf = N $d\Phi/dt$

- Example: for uniform **B**=2 sin(t) Tesla/s as shown above in a loop of 2 m radius, find the emf

$$\text{emf} = \frac{d\Phi}{dt} = \frac{\partial}{\partial t}\int_S \mathbf{B} \cdot d\mathbf{S} = \frac{\partial}{\partial t}\int_S 2\sin(t)\hat{\mathbf{z}} \cdot dS\,\hat{\mathbf{z}} = \frac{\partial}{\partial t}2\pi R^2 \sin(t) = 8\pi \cos(t)$$

Emf in an Inductor

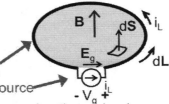

$$\text{emf} = -\frac{d\Phi}{dt} = -\frac{\partial}{\partial t}\int_S \mathbf{B} \cdot d\mathbf{S} = \oint_L \mathbf{E} \cdot d\mathbf{L}$$

perfect conductor wire loop
gap in wire loop with current source

- Consider the inductor above, where a conducting wire loop having a small gap of width Δ is driven with a current source i_L that generates a flux Φ and B-field **B** as shown

- For an emf line integral through the wire loop, where **E**=0 inside a conductor, the entirety of the emf appears across the gap as emf = $\int\mathbf{E}\cdot d\mathbf{L} \approx E_g \Delta$, where E_g is the E-field in the gap in the direction of d**L**. Also, V_g is
$$\text{gap voltage } V_g = V_{g+} - V_{g-} = -\int_{g-}^{g+} \mathbf{E} \cdot d\mathbf{L} \approx -E_g\Delta = -emf$$

- Substituting emf= $-d\Phi/dt$, then V_g= -emf = $d\Phi/dt$

- For a 1-turn inductor L= $N\Phi/i_L$ = Φ/i_L (or $i=\Phi/L$) and v_L= L di_L/dt

- Substuituting, v_L= L di_L/dt= L $d(\Phi/L)/dt$ = $d\Phi/dt$ = -emf =V_g

- So, the inductor voltage is v_L= V_g = -emf

Emf in a Multi-Loop Inductor

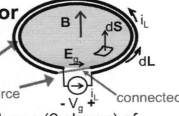

$$\text{emf} = -\frac{d\Phi}{dt} = -\frac{\partial}{\partial t}\int_S \mathbf{B} \cdot d\mathbf{S} = \oint_L \mathbf{E} \cdot d\mathbf{L}$$

2 wire loops with infinitesimal separation
gap in wire loop with current source
connected

- Consider the previous case, but with N loops (2 shown) of perfect conductor, connected as shown with current source i_L, that now creates flux Φ_T= N Φ_1, N times the flux Φ_1 of one loop

- If the emf line integral for each loop would be emf$_1$ = $\int_{L1}\mathbf{E}\cdot d\mathbf{L} \approx$ $E_g \Delta$ =$-d\Phi_T/dt$, for each loop if they were not connected. So, when connected in series, emf = N emf$_1$= -N $d\Phi_T/dt$.

- As before, V_g= -emf = N $d\Phi_T/dt$

- Recall for an inductor L=$N\Phi_T/i_L$ (or $i=N\Phi_T/L$) and v_L= L di_L/dt

- Substituting, v_L= L di_L/dt= L $d(N\Phi_T/L)/dt$ = N $d\Phi_T/dt$ = V_g

- So, the inductor voltage is v_L= V_g = -emf

- And L_{nturn}= $N\Phi_T/i_L$ =N(NΦ_1)/i_L =$N^2(\Phi_1/i_L)$ = $N^2 L_{1turn}$

Emf in an Open Transformer

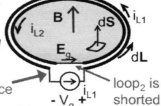

$$\text{emf} = -\frac{d\Phi}{dt} = -\frac{\partial}{\partial t}\int_{S}\mathbf{B}\cdot d\mathbf{S} = \oint_{L}\mathbf{E}\cdot d\mathbf{L}$$

2 wire loops with infinitesimal separation

gap in wire loop with current source

loops not connected

- Consider the previous case, but with 2 unconnected wire loops, and with current source i_L in loop$_1$ that now creates flux Φ_1 shared by both loops. For identical loops, inductance $L_1=L_2=L$
- Since both loops share the same flux, emf$_1$= emf$_2$= -$d\Phi_1$/dt
- Emf in loop$_1$ is emf$_1$= $\int_{L1}\mathbf{E}\cdot d\mathbf{L} \approx \mathbf{E}_g \Delta$, and as before V_{g1}= -emf
- Since emf$_1$=emf$_2$, then V_{g2}=V_{g1}= -emf$_1$= $d\Phi_1$/dt
- For the loop$_1$ inductor $L= \Phi_1/i_{L1}$ (or $i_1= \Phi_1/L$) and v_{L1}= L di_{L1}/dt
- Substituting, v_{L1}= L di_{L1}/dt= L d(Φ_1/L)/dt = $d\Phi_1$/dt = V_{g1}
- So, loop$_1$ and loop$_2$ inductor voltages are v_{L1}= v_{L2}= V_{g1} = $d\Phi_1$/dt
- Since loop$_2$ is has a gap, i_{L2}=0, so flux from i_{L2} in loop$_2$ is Φ_2=0
- For this transformer: v_{L1}= v_{L2} =L di_{L1}/dt, and i_{L2}=0

Emf in an Shorted Transformer

$$\text{emf} = -\frac{d\Phi}{dt} = -\frac{\partial}{\partial t}\int_{S}\mathbf{B}\cdot d\mathbf{S} = \oint_{L}\mathbf{E}\cdot d\mathbf{L}$$

2 wire loops with infinitesimal separation

gap in wire loop with current source

loop$_2$ is shorted

- Consider the case with loop$_2$ short-circuited in the gap, for 2 loops unconnected, and current source in loop$_1$ sets i_{L1} that creates flux Φ_1 shared by the loops. Again, inductance $L_1=L_2=L$.
- Both loops share same total flux Φ_T, so emf$_1$= emf$_2$= -$d\Phi_T$/dt
- Because of the short circuit in loop$_2$, emf$_2$= $\int_{loop2}\mathbf{E}\cdot d\mathbf{L}$ =0
- Since emf$_1$=emf$_2$ as before, V_{g1}= -emf$_1$= -emf$_2$ = V_{g2}= 0
- Since 0=emf$_1$= -$d\Phi_T$/dt = -d(Φ_1+Φ_2)/dt, and $d\Phi_1$/dt= -$d\Phi_2$/dt
- For identical loops, $L_1=L_2=L$ so (Φ_1+Φ_2)=L (i_{L1}+i_{L2})
- Then 0=d(Φ_1+Φ_2)/dt= L (di_{L1}/dt + di_{L2}/dt), so di_{L1}/dt = -di_{L2}/dt
- Integrating, i_{L2} = -i_{L1} + C, where C is a constant
- For this transformer: v_{L1}= v_{L2}=0 and i_{L2}= -i_{L1} + C

Lenz's Law

- Lenz's law roughly states that currents that are induced by a changing external magnetic field tend to act in such a way as to oppose and reduce the external magnetic field

- In the foregoing case, we have seen an extreme example of Lenz's law, where the short-circuited winding created a changing magnetic flux $d\Phi_2/dt$ that completely canceled the changing magnetic flux $d\Phi_1/dt$ of the other winding, such that the total changing magnetic flux $d\Phi_T/dt=0$

- More typically, less drastic reductions in the changing magnetic field are observed

- A common situation where this can affect electronic circuits is when inductors are near ground planes and other metal objects

- Since $L=\Phi/i$, reduced flux can result in reduced inductance

Resistor-Loaded Transformer

$$\text{emf} = -\frac{d\Phi}{dt} = -\frac{\partial}{\partial t}\int_S \mathbf{B}\cdot d\mathbf{S} = \oint_L \mathbf{E}\cdot d\mathbf{L}$$

2 wire loops with infinitesimal separation

gap in wire loop with current source

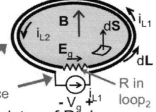

- Consider the previous case, but with a resistor of R ohms connected in the gap of $loop_2$, loops unconnected, and current source in $loop_1$ sets current i_{L1} that creates flux Φ_1 in both loops

- Both loops share same total flux Φ_T, so $\text{emf}_1 = \text{emf}_2 = -d\Phi_T/dt$

- As before, $V_{g1} = -\text{emf}_1 = -\text{emf}_2 = V_{g2}$, then $i_{L2} = -V_{g2}/R$

- As before, $L_1 = L_2 = L$, so $L_1 = \Phi_1/i_{L1}$ and $L_2 = \Phi_2/i_{L2}$

- In different form, $\Phi_1 = L\, i_{L1}$ and $\Phi_2 = L\, i_{L2}$

- Adding together: $\Phi_1 + \Phi_2 = L\, i_{L1} + L\, i_{L2} = L(i_{L1} + i_{L2})$

- If $L \gg \Phi_T$, $d(i_{L1} + i_{L2})/dt = d(\Phi_1 + \Phi_2)/dt/L = d\Phi_T/dt/L \approx 0$

- So $i_{L1} \approx -i_{L2} + C = V_{g2}/R + C$ for large inductance, so $L \gg \Phi_T$

- For this transformer: $v_{L1} = v_{L2} = V_{g2} = i_{L1}R$ and $i_{L2} \approx -i_{L1}$

Point Form of Faraday's Law

- Recall Stokes' theorem:

$$\int_S \nabla \times \mathbf{A} \cdot d\mathbf{S} = \oint_L \mathbf{A} \cdot d\mathbf{L}$$

- Apply Stokes' theorem to the line integral in Faraday's Law

$$\text{emf} = -\frac{d\Phi}{dt} = -\frac{\partial}{\partial t}\int_S \mathbf{B} \cdot d\mathbf{S} = \oint_L \mathbf{E} \cdot d\mathbf{L} = \int_S \nabla \times \mathbf{E} \cdot d\mathbf{S}$$

so for a fixed surface S, then

$$-\int_S \frac{\partial \mathbf{B}}{\partial t} \cdot d\mathbf{S} = \int_S \nabla \times \mathbf{E} \cdot d\mathbf{S} \quad \text{and so} \quad \Rightarrow \quad \nabla \times \mathbf{E} = -\frac{\partial \mathbf{B}}{\partial t}$$

- Finally, the point form and integral form of Faraday's Law:

Point Form (Differential form)   Integral Form

$$\nabla \times \mathbf{E} = -\frac{\partial \mathbf{B}}{\partial t} \quad \text{Faraday's Law} \qquad \text{emf} = -\frac{d\Phi}{dt} = -\frac{\partial}{\partial t}\int_S \mathbf{B} \cdot d\mathbf{S} = \oint_L \mathbf{E} \cdot d\mathbf{L}$$

- This is the first law where we see interaction between electric and magnetic fields in point form

Displacement Current

The Need for Displacement Current

- Recall Ampere's circuital law for magnetostatics

Magnetostatic Point Form

$$\nabla \times \mathbf{H} = \mathbf{J}_e$$

Ampere Circuital Law

Magnetostatic Integral Form

$$\oint_L \mathbf{H} \cdot d\mathbf{L} = \int_S \mathbf{J}_e \cdot d\mathbf{S} = I$$

- The point-form $\nabla \times \mathbf{H} = \mathbf{J}_e$ is similar to the point form of Faraday, except that there is no time derivative on the right side

- To investigate, take the divergence of both sides:

$$\nabla \cdot (\nabla \times \mathbf{H}) = 0 = \nabla \cdot \mathbf{J}_e \quad ???$$

- The zero above follows from using a vector identity on the left

- However, this is inconsistent with our earlier point continuity

$$\nabla \cdot \mathbf{J} = -\frac{\partial}{\partial t} \rho_e = 0?? \quad \text{no, does not necessarily equal zero}$$

- The solution by Maxwell is to add displacement current d$\mathbf{D}$/dt

$$\nabla \times \mathbf{H} = \mathbf{J}_e + \frac{\partial \mathbf{D}}{\partial t}$$

- Then everything becomes consistent

$$\nabla \cdot (\nabla \times \mathbf{H}) = \nabla \cdot \mathbf{J}_e + \frac{\partial \nabla \cdot \mathbf{D}}{\partial t} = -\frac{\partial}{\partial t} \rho_e + \frac{\partial}{\partial t} \rho_e = 0$$

Ampere's Law with Displacement Current

- Next, we find the integral form by first integrating over a surface, similar to the current/charge conservation derivation except the current is now through an open surface

$$I_{through} = \int_S (\nabla \times \mathbf{H}) \cdot d\mathbf{S} = \int_S \left(\mathbf{J}_e + \frac{\partial \mathbf{D}}{\partial t} \right) \cdot d\mathbf{S}$$

- Recall Stoke's thoerem

$$\int_S \nabla \times \mathbf{A} \cdot d\mathbf{S} = \oint_L \mathbf{A} \cdot d\mathbf{L}$$

- Substituting

$$I_{through} = \int_S (\nabla \times \mathbf{H}) \cdot d\mathbf{S} = \oint_L \mathbf{H} \cdot d\mathbf{L} = \int_S \left(\mathbf{J}_e + \frac{\partial \mathbf{D}}{\partial t} \right) \cdot d\mathbf{S}$$

- Finally, point and integral forms of Ampere's Circuital Law are

  Point Form

  $$\nabla \times \mathbf{H} = \mathbf{J}_e + \frac{\partial \mathbf{D}}{\partial t}$$

  Ampere Circuital Law

  Integral Form

  $$\oint_L \mathbf{H} \cdot d\mathbf{L} = \int_S \left(\mathbf{J}_e + \frac{\partial \mathbf{D}}{\partial t} \right) \cdot d\mathbf{S} = I$$

- This is the second law where we see interaction between electric and magnetic fields in point form

Applying Ampere's Circuital Law

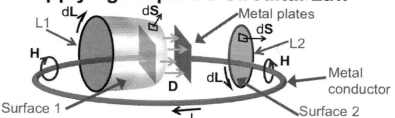

- The device above consists of a wire with current I connected to two metal plates with a D-field between. The wire passes through both open surfaces, with perimeters L1 and L2

- At surface 1:
  - The line integral follows perimeter L1 with d**L** and d**S** shown
  - Since L1 surrounds the wire, $\int_{L1}$**H**·d**L** = I = wire current
  - No conduction current crosses surface 1, so **J**$_e$ =0, then:

$$I = \oint_L \mathbf{H} \cdot d\mathbf{L} = \int_S \left(\mathbf{J_e} + \frac{\partial \mathbf{D}}{\partial t} \right) \cdot d\mathbf{S} = \int_S \frac{\partial \mathbf{D}}{\partial t} \cdot d\mathbf{S}$$

  Note: $\int d\mathbf{D}/dt \cdot d\mathbf{S}$ equals current I in the wire

  - Only displacement current crosses surface 1

Applying Ampere's Circuital Law

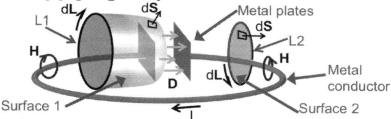

- The device above consists of a wire with current I connected to two metal plates with a D-field between. The wire passes through both open surfaces, with perimeters L1 and L2

- At surface 2:
  - The line integral follows perimeter L2 with d**L** and d**S** shown
  - Since L2 surrounds the wire, $\int_{L2}$**H**·d**L** = I = wire current
  - No displacement current crosses surface 1, so **D** =0, then:

$$I = \oint_L \mathbf{H} \cdot d\mathbf{L} = \int_S \left(\mathbf{J_e} + \frac{\partial \mathbf{D}}{\partial t} \right) \cdot d\mathbf{S} = \int_S \mathbf{J_e} \cdot d\mathbf{S}$$

  Note: $\int \mathbf{J_e} \cdot d\mathbf{S}$ equals current I in the wire

  - Only conduction current crosses surface 2

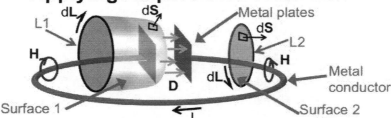

Applying Ampere's Circuital Law

- Example: let current I be I(t)= t u(t), and two metal plates be 1x1 cm, 1 mm apart. Find D(t) if D(0)=0, and find dD(t)/dt
- At surface 1:
  - Since L1 surrounds the wire, $\int_{L1} \mathbf{H} \cdot d\mathbf{L} = I = t\, u(t)$
  - No conduction current crosses surface 1, so $\mathbf{J_e} = 0$, then:

$$I = t\,u(t) = \oint_L \mathbf{H} \cdot d\mathbf{L} = \int_S \left(\mathbf{J_e} + \frac{\partial \mathbf{D}}{\partial t} \right) \cdot d\mathbf{S} = \int_S \frac{\partial \mathbf{D}}{\partial t} \cdot d\mathbf{S} \approx (0.01)^2 \frac{\partial D(t)}{\partial t}$$

$$\text{so} \quad D(t) = \frac{1}{(0.01)^2} \int t\,u(t)\,dt = \frac{1}{(0.01)^2} \frac{t^2}{2} u(t) + k; \ D(t) = 0, so\ k = 0$$

$$D(t) = 5000\,t^2\,u(t)\,\text{C/m}^2, \quad \partial D(t)/\partial t = 10,000\,t\,u(t)\,\text{A/m}^2,$$

Einstein

- Einstein's comment on a bit of inconsistency in Faraday's Law motivates relativity

"The observable phenomenon here depends only on the relative motion of the conductor and the magnet, whereas the customary view draws a sharp distinction between the two cases in which either the one or the other of these bodies is in motion. For if the magnet is in motion and the conductor at rest, there arises in the neighbourhood of the magnet an electric field with a certain definite energy, producing a current at the places where parts of the conductor are situated.

But if the magnet is stationary and the conductor in motion, no electric field arises in the neighbourhood of the magnet."

Albert Einstein, On the Electrodynamics of Moving Bodies.

Time-Varying Maxwell's Equations

Putting Everything Together

- The new time varying results can now be used to update the Maxwell's equations from their static forms
- Summarizing:
  - o New equations from Faraday's law

Point Form (Differential form) Integral Form

$$\nabla \times \mathbf{E} = -\frac{\partial \mathbf{B}}{\partial t} \quad \text{Faraday's Law} \quad \text{emf} = -\frac{d\Phi}{dt} = -\frac{\partial}{\partial t} \int_S \mathbf{B} \cdot d\mathbf{S} = \oint_L \mathbf{E} \cdot d\mathbf{L}$$

  - o New equations from Ampere's circuital law

Point Form Integral Form

$$\nabla \times \mathbf{H} = \mathbf{J}_e + \frac{\partial \mathbf{D}}{\partial t} \quad \text{Ampere Circuital Law} \quad \oint_L \mathbf{H} \cdot d\mathbf{L} = \int_S \left(\mathbf{J}_e + \frac{\partial \mathbf{D}}{\partial t} \right) \cdot d\mathbf{S} = I$$

- The remaining Maxwell relations are unchanged

Maxwell's Equations

Point Form (Differential form)   Integral Form

$$\nabla \times \mathbf{E}(\overline{x},t) = -\frac{\partial \mathbf{B}(\overline{x},t)}{\partial t}$$ Faraday's Law $$\oint_L \mathbf{E}(\overline{x},t) \cdot d\mathbf{L} = -\int_S \frac{\partial \mathbf{B}(\overline{x},t)}{\partial t} \cdot d\mathbf{S}$$

assuming $\mathbf{J}_m(\overline{x},t) = 0$

$$\nabla \times \mathbf{H}(\overline{x},t) = \mathbf{J}_e(\overline{x},t) + \frac{\partial \mathbf{D}(\overline{x},t)}{\partial t}$$ Ampere Circuital Law $$\oint_L \mathbf{H}(\overline{x},t) \cdot d\mathbf{L} = \int_S \left(\mathbf{J}_e(\overline{x},t) + \frac{\partial \mathbf{D}(\overline{x},t)}{\partial t} \right) \cdot d\mathbf{S}$$

$$\nabla \cdot \mathbf{D}(\overline{x},t) = \rho_e(\overline{x},t)$$ Gauss' Law $$Q = \int_V \rho_e(\overline{x},t) \, dv = \oint_S \mathbf{D}(\overline{x},t) \cdot d\mathbf{S}$$

$$\nabla \cdot \mathbf{B}(\overline{x},t) = 0$$ Gauss' Magnetism Law $$0 = \oint_S \mathbf{B}(\overline{x},t) \cdot d\mathbf{S}$$

assuming $\rho_m(\overline{x},t) = 0$

$$\mathbf{D}(\overline{x},t) = \varepsilon \mathbf{E}(\overline{x},t) \quad \text{if } \varepsilon(t) = \delta(t)\varepsilon$$ $$\mathbf{J}_e = \sigma \mathbf{E}$$

$$\mathbf{B}(\overline{x},t) = \mu \mathbf{H}(\overline{x},t) \quad \text{if } \mu(t) = \delta(t)\mu$$ and $\overline{x} = [x \; y \; z]^T$

otherwise

$$\mathbf{D}(\overline{x},t) = \varepsilon(t) * \mathbf{E}(\overline{x},t)$$

$$\mathbf{B}(\overline{x},t) = \mu(t) * \mathbf{H}(\overline{x},t)$$

Maxwell's Equations (Shorthand)

Point Form (Differential form)   Integral Form

$$\nabla \times \mathbf{E} = -\frac{\partial \mathbf{B}}{\partial t}$$ Faraday's Law $$\oint_L \mathbf{E} \cdot d\mathbf{L} = -\int_S \frac{\partial \mathbf{B}}{\partial t} \cdot d\mathbf{S}$$

assuming $\mathbf{J}_m = 0$

$$\nabla \times \mathbf{H} = \mathbf{J}_e + \frac{\partial \mathbf{D}}{\partial t}$$ Ampere Circuital Law $$\oint_L \mathbf{H} \cdot d\mathbf{L} = \int_S \left(\mathbf{J}_e + \frac{\partial \mathbf{D}}{\partial t} \right) \cdot d\mathbf{S}$$

$$\nabla \cdot \mathbf{D} = \rho_e$$ Gauss' Law $$Q = \int_V \rho_e \, dv = \oint_S \mathbf{D} \cdot d\mathbf{S}$$

$$\nabla \cdot \mathbf{B} = 0$$ Gauss' Magnetism Law $$0 = \oint_S \mathbf{B} \cdot d\mathbf{S}$$

assuming $\rho_m(\overline{x},t) = 0$

$$\mathbf{D} = \varepsilon \mathbf{E} \quad \text{if } \varepsilon(t) = \delta(t)\varepsilon$$ $$\mathbf{J}_e = \sigma \mathbf{E}$$

$$\mathbf{B} = \mu \mathbf{H} \quad \text{if } \mu(t) = \delta(t)\mu$$ and $\overline{x} = [x \; y \; z]^T$

otherwise

$$\mathbf{D} = \varepsilon(t) * \mathbf{E}$$

$$\mathbf{B} = \mu(t) * \mathbf{H}$$

6 MAXWELL'S EQUATIONS AND WAVES

The lecture notes in this chapter present Maxwell's equations and electromagnetic waves
.

Phasor Form of Maxwell's Equations

Using Phasors in Maxwell's Equations

- Before proceeding to solve Maxwell's equations for wave solutions, we quickly review phasors and develop the phasor form of Maxwell's equations

- Recall that phasors have an "implied" $e^{j\omega t}$ term

$V_0 e^{-\gamma x} e^{j\omega t} = |V_0| e^{j\theta} e^{-\gamma x} e^{j\omega t}$ in phasor form is: $V_0 e^{-\gamma x}$

if $\gamma = \alpha + j\beta,$ $\text{Re}\left\{ |V_0| e^{j\theta} e^{-\alpha x} e^{-j\beta x} e^{j\omega t} \right\} = |V_0| e^{-\alpha x} \cos(\omega t - \beta x + \theta)$

- And phasors give simple solutions to differential equations

if $\dfrac{\partial^2 v(x,t)}{\partial x^2} = 9 \dfrac{\partial^2 v(x,t)}{\partial t^2}$ guess the solution is: $Ae^{-\gamma x} e^{j\omega t}$

$\Rightarrow \dfrac{\partial^2 Ae^{-\gamma x} e^{j\omega t}}{\partial x^2} = 9 \dfrac{\partial^2 Ae^{-\gamma x} e^{j\omega t}}{\partial t^2} \Rightarrow \gamma^2 \left(Ae^{-\gamma x} e^{j\omega t} \right) = -9\omega^2 \left(Ae^{-\gamma x} e^{j\omega t} \right)$

$\Rightarrow$ then $\gamma^2 = -9\omega^2$ or $\gamma = \pm j3\omega,$ so the solution is: $Ae^{\pm j3\omega x} e^{j\omega t}$

in phasor notation, the solution is: $Ae^{\pm j3\omega x}$

- Phasors not needed in statics with no time variation

Maxwell's Equations

Point Form (Differential form)   Integral Form

$$\nabla \times \mathbf{E}(\overline{x},t) = -\frac{\partial \mathbf{B}(\overline{x},t)}{\partial t}$$ Faraday's Law $$\oint_L \mathbf{E}(\overline{x},t) \cdot d\mathbf{L} = -\int_S \frac{\partial \mathbf{B}(\overline{x},t)}{\partial t} \cdot d\mathbf{S}$$

assuming $\mathbf{J}_m(\overline{x},t) = 0$

$$\nabla \times \mathbf{H}(\overline{x},t) = \mathbf{J}_e(\overline{x},t) + \frac{\partial \mathbf{D}(\overline{x},t)}{\partial t}$$ Ampere Circuital Law $$\oint_L \mathbf{H}(\overline{x},t) \cdot d\mathbf{L} = \int_S \left(\mathbf{J}_e(\overline{x},t) + \frac{\partial \mathbf{D}(\overline{x},t)}{\partial t} \right) \cdot d\mathbf{S}$$

$$\nabla \cdot \mathbf{D}(\overline{x},t) = \rho_e(\overline{x},t)$$ Gauss' Law $$Q = \int_V \rho_e(\overline{x},t)\,dv = \oint_S \mathbf{D}(\overline{x},t) \cdot d\mathbf{S}$$

$$\nabla \cdot \mathbf{B}(\overline{x},t) = 0$$ Gauss' Magnetism Law $$0 = \oint_S \mathbf{B}(\overline{x},t) \cdot d\mathbf{S}$$

assuming $\rho_m(\overline{x},t) = 0$

$\mathbf{D}(\overline{x},t) = \varepsilon \mathbf{E}(\overline{x},t)$ if $\varepsilon(t) = \delta(t)\varepsilon$ $\mathbf{J}_e = \sigma \mathbf{E}$

$\mathbf{B}(\overline{x},t) = \mu \mathbf{H}(\overline{x},t)$ if $\mu(t) = \delta(t)\mu$ and $\overline{x} = [x\ y\ z]^T$

otherwise

$\mathbf{D}(\overline{x},t) = \varepsilon(t) * \mathbf{E}(\overline{x},t)$

$\mathbf{B}(\overline{x},t) = \mu(t) * \mathbf{H}(\overline{x},t)$

Maxwell's Equations (Shorthand)

Point Form (Differential form)   Integral Form

$$\nabla \times \mathbf{E} = -\frac{\partial \mathbf{B}}{\partial t}$$ Faraday's Law $$\oint_L \mathbf{E} \cdot d\mathbf{L} = -\int_S \frac{\partial \mathbf{B}}{\partial t} \cdot d\mathbf{S}$$

assuming $\mathbf{J}_m = 0$

$$\nabla \times \mathbf{H} = \mathbf{J}_e + \frac{\partial \mathbf{D}}{\partial t}$$ Ampere Circuital Law $$\oint_L \mathbf{H} \cdot d\mathbf{L} = \int_S \left(\mathbf{J}_e + \frac{\partial \mathbf{D}}{\partial t} \right) \cdot d\mathbf{S}$$

$$\nabla \cdot \mathbf{D} = \rho_e$$ Gauss' Law $$Q = \int_V \rho_e\,dv = \oint_S \mathbf{D} \cdot d\mathbf{S}$$

$$\nabla \cdot \mathbf{B} = 0$$ Gauss' Magnetism Law $$0 = \oint_S \mathbf{B} \cdot d\mathbf{S}$$

assuming $\rho_m(\overline{x},t) = 0$

$\mathbf{D} = \varepsilon \mathbf{E}$ if $\varepsilon(t) = \delta(t)\varepsilon$ $\mathbf{J}_e = \sigma \mathbf{E}$

$\mathbf{B} = \mu \mathbf{H}$ if $\mu(t) = \delta(t)\mu$ and $\overline{x} = [x\ y\ z]^T$

otherwise

$\mathbf{D} = \varepsilon(t) * \mathbf{E}$

$\mathbf{B} = \mu(t) * \mathbf{H}$

Phasor Form of Faraday's Law

- To find the phasor form of Faraday's law, we could substitute the generic vector form "$\mathbf{A}\, e^{-\gamma z}\, e^{j\omega t}$" for each vector,
- However, a more general phasor is "$\mathbf{A}(x,y,z)\, e^{j\omega t}$", then:

Point Form (Differential form)   Integral Form

$$\nabla \times \mathbf{E} = -\frac{\partial \mathbf{B}}{\partial t} \quad \text{Faraday's Law} \quad emf = -\frac{d\Phi}{dt} = -\frac{\partial}{\partial t}\int_S \mathbf{B}\cdot d\mathbf{S} = \oint_L \mathbf{E}\cdot d\mathbf{L}$$

$$\Rightarrow \nabla \times \mathbf{E}(x,y,z)e^{j\omega t} = -\frac{\partial \mathbf{B}(x,y,z)e^{j\omega t}}{\partial t} = -j\omega\, \mathbf{B}(x,y,z)e^{j\omega t}$$

$$\Rightarrow \oint_L \mathbf{E}(x,y,z)e^{j\omega t}\cdot d\mathbf{L} = -\frac{\partial}{\partial t}\int_S \mathbf{B}(x,y,z)e^{j\omega t}\cdot d\mathbf{S} = -j\omega\int_S \mathbf{B}(x,y,z)e^{j\omega t}\cdot d\mathbf{S}$$

- Dropping the "$(x,y,z)\, e^{j\omega t}$", phasor form of Faraday's law is:

Point Form (Differential form)   Integral Form

$$\nabla \times \mathbf{E} = -j\omega\, \mathbf{B} \qquad\qquad \oint_L \mathbf{E}\cdot d\mathbf{L} = -j\omega\int_S \mathbf{B}\cdot d\mathbf{S}$$

Phasor Form of Ampere's Circuital Law

- To find the phasor form of Ampere's law, we do the same as for Faraday, using phasor "$\mathbf{A}(x,y,z)\, e^{j\omega t}$", then:

Point Form (Differential form)   Integral Form

$$\nabla \times \mathbf{H} = \mathbf{J}_e + \frac{\partial \mathbf{D}}{\partial t} \quad \text{Ampere Circ. Law} \quad \oint_L \mathbf{H}\cdot d\mathbf{L} = \int_S \left(\mathbf{J}_e + \frac{\partial \mathbf{D}}{\partial t}\right)\cdot d\mathbf{S} = I$$

$$\Rightarrow \nabla \times \mathbf{H}(x,y,z)e^{j\omega t} = \mathbf{J}_e(x,y,z)e^{j\omega t} + \frac{\partial \mathbf{D}(x,y,z)e^{j\omega t}}{\partial t}$$

$$= \mathbf{J}_e(x,y,z)e^{j\omega t} + j\omega\, \mathbf{D}(x,y,z)e^{j\omega t}$$

$$\Rightarrow \oint_L \mathbf{H}\cdot d\mathbf{L} = \int_S \left(\mathbf{J}_e + \frac{\partial \mathbf{D}}{\partial t}\right)\cdot d\mathbf{S} = \int_S \left(\mathbf{J}_e + j\omega\mathbf{D}\right)\cdot d\mathbf{S}$$

- Dropping the "$(x,y,z)\, e^{j\omega t}$", phasor form of Ampere's law is:

Point Form (Differential form)   Integral Form

$$\nabla \times \mathbf{H} = \mathbf{J}_e + j\omega\, \mathbf{D} \qquad\qquad \oint_L \mathbf{H}\cdot d\mathbf{L} = \int_S \left(\mathbf{J}_e + j\omega\mathbf{D}\right)\cdot d\mathbf{S}$$

Phasor Form of Remaining Point Forms

- The remaining equations do not involve derivatives, and so transform directly into phasor form

Original forms

Phasor Forms

$$\nabla \times \mathbf{E}(\bar{x},t) = -\frac{\partial \mathbf{B}(\bar{x},t)}{\partial t} \qquad \Rightarrow \qquad \nabla \times \mathbf{E} = -j\omega \mathbf{B}$$

assuming $\mathbf{J}_m(\bar{x},t) = 0$

$$\nabla \times \mathbf{H}(\bar{x},t) = \mathbf{J}_e(\bar{x},t) + \frac{\partial \mathbf{D}(\bar{x},t)}{\partial t} \qquad \Rightarrow \qquad \nabla \times \mathbf{H} = \mathbf{J}_e + j\omega \mathbf{D}$$

$$\nabla \cdot \mathbf{D}(\bar{x},t) = \rho_e(\bar{x},t) \qquad \Rightarrow \qquad \nabla \cdot \mathbf{D} = \rho_e$$

$$\nabla \cdot \mathbf{B}(\bar{x},t) = 0 \qquad \Rightarrow \qquad \nabla \cdot \mathbf{B} = 0$$

assuming $\rho_m(\bar{x},t) = 0$

$$\mathbf{D}(\bar{x},t) = \varepsilon \mathbf{E}(\bar{x},t) \qquad \Rightarrow \qquad \mathbf{D} = \varepsilon \mathbf{E}$$

$$\mathbf{B}(\bar{x},t) = \mu \mathbf{H}(\bar{x},t) \qquad \Rightarrow \qquad \mathbf{B} = \mu \mathbf{H}$$

where $\bar{x} = \begin{bmatrix} x & y & z \end{bmatrix}^T$

Phasor Form of Maxwell's Equations

- The point and integral phasor forms

Point Form (Differential form)

Integral Form

$$\nabla \times \mathbf{E} = -j\omega \mathbf{B} \qquad \text{Faraday's Law} \qquad \oint_L \mathbf{E} \cdot d\mathbf{L} = -j\omega \int_S \mathbf{B} \cdot d\mathbf{S}$$

assuming $\mathbf{J}_m = 0$

$$\nabla \times \mathbf{H} = \mathbf{J}_e + j\omega \mathbf{D} \qquad \begin{array}{c} \text{Ampere} \\ \text{Circuital Law} \end{array} \qquad \oint_L \mathbf{H} \cdot d\mathbf{L} = \int_S \left(\mathbf{J}_e + j\omega \mathbf{D} \right) \cdot d\mathbf{S}$$

$$\nabla \cdot \mathbf{D} = \rho_e \qquad \text{Gauss' Law} \qquad Q = \int_V \rho_e \, dv = \oint_S \mathbf{D} \cdot d\mathbf{S}$$

$$\nabla \cdot \mathbf{B} = 0 \qquad \text{Gauss' Magnetism Law} \qquad 0 = \oint_S \mathbf{B} \cdot d\mathbf{S}$$

assuming $\rho_m(\bar{x},t) = 0$

$$\mathbf{D} = \varepsilon \mathbf{E} \qquad \qquad \mathbf{J}_e = \sigma \mathbf{E}$$

$$\mathbf{B} = \mu \mathbf{H}$$

Caution on Maxwell's Equations

- Maxwell's Equations
  - ○ Often μ and ε are presumed to be constant:

$$\nabla \times E = -\frac{\partial}{\partial t}B - J_m \qquad \cancel{B = \mu H}$$

$$\nabla \times H = \frac{\partial}{\partial t}D + J \qquad \cancel{D = \varepsilon E} \qquad J = \sigma E$$

- More precisely:

$$B(t) = \mu(t) * H(t) \quad so \quad B(\omega) = \mu(\omega)H(\omega)$$
$$D(t) = \varepsilon(t) * E(t) \quad so \quad D(\omega) = \varepsilon(\omega)E(\omega)$$

Wave Equation Obtained from Maxwell's Equations

Wave Equation from Maxwell's Equations

- Just as the Telegrapher's equations were combined to find the wave equation for RC-transmission lines, Maxwell's equations yield the wave equation for wave propagation in space

- To begin, start with the Faraday and Ampere point forms:

  Faraday: $\nabla \times \mathbf{E} = -\dfrac{\partial \mathbf{B}}{\partial t}$ Ampere: $\nabla \times \mathbf{H} = \mathbf{J}_e + \dfrac{\partial \mathbf{D}}{\partial t}$

- Taking the curl of Faraday, and several further steps

Faraday: $\nabla \times \nabla \times \mathbf{E} = -\nabla \times \dfrac{\partial \mathbf{B}}{\partial t} = -\dfrac{\partial \nabla \times \mathbf{B}}{\partial t} = -\mu \dfrac{\partial \nabla \times \mathbf{H}}{\partial t}$ for $\mathbf{B} = \mu \mathbf{H}$

then $\nabla(\nabla \cdot \mathbf{E}) - \nabla^2 \mathbf{E} = -\mu \dfrac{\partial \nabla \times \mathbf{H}}{\partial t}$ because $\nabla \times \nabla \times \mathbf{E} = \nabla(\nabla \cdot \mathbf{E}) - \nabla^2 \mathbf{E}$

then $\nabla^2 \mathbf{E} = \mu \dfrac{\partial \nabla \times \mathbf{H}}{\partial t}$ because $(\nabla \cdot \mathbf{E}) = 0$ in source-free region

finally $\boxed{\nabla^2 \mathbf{E} = \mu \varepsilon \dfrac{\partial^2 \mathbf{E}}{\partial t^2}}$ substituting source-free $\mathbf{J}_e = 0$ Ampere law

- This is the wave equation from Maxwell's equations

Wave Equation in H

- The same procedure can be used to find the wave equation in terms of the H-field

- Again, start with the Faraday and Ampere point forms:

  Faraday: $\nabla \times \mathbf{E} = -\dfrac{\partial \mathbf{B}}{\partial t}$ Ampere: $\nabla \times \mathbf{H} = \mathbf{J}_e + \dfrac{\partial \mathbf{D}}{\partial t}$

- Now take the curl of Ampere, and several further steps

Ampere: $\nabla \times \nabla \times \mathbf{H} = -\nabla \times \dfrac{\partial \mathbf{D}}{\partial t} = -\dfrac{\partial \nabla \times \mathbf{D}}{\partial t} = -\varepsilon \dfrac{\partial \nabla \times \mathbf{E}}{\partial t}$ for $\mathbf{D} = \mu \varepsilon \mathbf{E}$

then $\nabla(\nabla \cdot \mathbf{H}) - \nabla^2 \mathbf{H} = -\varepsilon \dfrac{\partial \nabla \times \mathbf{E}}{\partial t}$ because $\nabla \times \nabla \times \mathbf{H} = \nabla(\nabla \cdot \mathbf{H}) - \nabla^2 \mathbf{H}$

then $\nabla^2 \mathbf{H} = \varepsilon \dfrac{\partial \nabla \times \mathbf{E}}{\partial t}$ because $(\nabla \cdot \mathbf{H}) = 0$

finally $\boxed{\nabla^2 \mathbf{H} = \mu \varepsilon \dfrac{\partial^2 \mathbf{H}}{\partial t^2}}$ substituting Faraday law

- This is the wave equation from Maxwell's equations

Wave Equations in Source-Free Region

- The wave equations are:

$$\nabla^2 \mathbf{E} = \mu\varepsilon\frac{\partial^2 \mathbf{E}}{\partial t^2} \qquad\qquad \nabla^2 \mathbf{H} = \mu\varepsilon\frac{\partial^2 \mathbf{H}}{\partial t^2}$$

- First, note the important assumptions in obtaining this wave equation
  - Must be a source-free region
    - With $\nabla\cdot\mathbf{D}=\rho_e=0$ meaning no free charge in the region
    - and $\mathbf{J}_e=0$ means no currents in the region
- Second, expand the Laplacian to see the details

$$\nabla^2 \mathbf{E} = \left(\frac{\partial^2}{\partial x^2}+\frac{\partial^2}{\partial y^2}+\frac{\partial^2}{\partial z^2}\right)\mathbf{E} = \mu\varepsilon\frac{\partial^2 \mathbf{E}}{\partial t^2}$$

- Which is quite similar to the RC-line wave equation:

$$\text{RC-line wave equation: } \frac{\partial^2 v(x,t)}{\partial x^2} = L_R C_R \frac{\partial^2 v(x,t)}{\partial t^2}$$

Helmholtz Equations in Source-Free Region

- The Helmholtz equations are the phasor form of the wave equations
- Given the wave equations:

$$\nabla^2 \mathbf{E} = \mu\varepsilon\frac{\partial^2 \mathbf{E}}{\partial t^2} \qquad\qquad \nabla^2 \mathbf{H} = \mu\varepsilon\frac{\partial^2 \mathbf{H}}{\partial t^2}$$

- The Helmholtz equations are:

$$\nabla^2 \mathbf{E} = -\omega^2\mu\varepsilon\mathbf{E} \qquad\qquad \nabla^2 \mathbf{H} = -\omega^2\mu\varepsilon\mathbf{H}$$

- With the usual phasor time convention of "A $e^{-\gamma x} e^{j\omega t}$" but now as a vector:

$$\text{phasor } \mathbf{E}e^{-\gamma z} \Rightarrow \mathbf{E}_0 e^{-\gamma z} e^{j\omega t} = \begin{bmatrix} E_{x0} \\ E_{y0} \\ E_{z0} \end{bmatrix} e^{-\gamma z} e^{j\omega t} = \begin{bmatrix} \left|E_{x0}\right|e^{j\theta_x} \\ \left|E_{y0}\right|e^{j\theta_y} \\ \left|E_{z0}\right|e^{j\theta_z} \end{bmatrix} e^{-\gamma z} e^{j\omega t}$$

- Helmholtz phasor forms often are easier to work with

Plane Wave Solution to Maxwell's Equations

$$\nabla^2 \mathbf{E} = -\mu\varepsilon \frac{\partial^2 \mathbf{E}}{\partial t^2}$$ **Plane Wave Solution Guess**

$$\nabla^2 \mathbf{E} = \left(\frac{\partial^2}{\partial x^2} + \frac{\partial^2}{\partial y^2} + \frac{\partial^2}{\partial z^2} \right) \mathbf{E} = -\mu\varepsilon \frac{\partial^2 \mathbf{E}}{\partial t^2}$$

- The plane-wave solution to the wave equation is found using 3 simplifying assumptions
  - The E-field vector only contains x-axis components
  - The wave only travels in z-axis direction (phase velocity v_p)
  - The region is source-free as before, $\nabla \cdot \mathbf{D} = \rho_e = 0$, $\mathbf{J_e} = 0$
- Guess that the solution has a plane wave similar to RC transmission line wave equation guess of "A $e^{-\gamma x} e^{j\omega t}$"
- Guess:

$$\text{guess: } \mathbf{E} = E_{x0} e^{-\gamma z} e^{j\omega t} \begin{bmatrix} 1 \\ 0 \\ 0 \end{bmatrix} = \begin{bmatrix} E_{x0} e^{-\gamma z} e^{j\omega t} \\ 0 \\ 0 \end{bmatrix}$$

Wave only travels in z-direction
since Re{$e^{-\gamma z} e^{j\omega t}$}=cos($\omega t - \gamma z$)

Only has x-component

Plane Wave Solution Steps

$$\nabla^2 \mathbf{E} = -\mu\varepsilon \frac{\partial^2 \mathbf{E}}{\partial t^2}$$

$$\nabla^2 \mathbf{E} = \left(\frac{\partial^2}{\partial x^2} + \frac{\partial^2}{\partial y^2} + \frac{\partial^2}{\partial z^2} \right) \mathbf{E} = -\mu\varepsilon \frac{\partial^2 \mathbf{E}}{\partial t^2}$$

- Using plane-wave solution guess from previous slide:

$$\left(\frac{\partial^2}{\partial x^2} + \frac{\partial^2}{\partial y^2} + \frac{\partial^2}{\partial z^2} \right) \begin{bmatrix} E_{x0} e^{-\gamma z} e^{j\omega t} \\ 0 \\ 0 \end{bmatrix} = \mu\varepsilon \frac{\partial^2 \mathbf{E}}{\partial t^2} = \mu\varepsilon \frac{\partial^2}{\partial t^2} \begin{bmatrix} E_{x0} e^{-\gamma z} e^{j\omega t} \\ 0 \\ 0 \end{bmatrix}$$

Only depends on z

$$\Rightarrow \frac{\partial^2}{\partial z^2} \begin{bmatrix} E_{x0} e^{-\gamma z} e^{j\omega t} \\ 0 \\ 0 \end{bmatrix} = \mu\varepsilon \frac{\partial^2}{\partial t^2} \begin{bmatrix} E_{x0} e^{-\gamma z} e^{j\omega t} \\ 0 \\ 0 \end{bmatrix}$$

$$\Rightarrow \frac{\partial^2 E_{x0} e^{-\gamma z} e^{j\omega t}}{\partial z^2} = \mu\varepsilon \frac{\partial^2 E_{x0} e^{-\gamma z} e^{j\omega t}}{\partial t^2}$$

$$\Rightarrow \gamma^2 E_{x0} e^{-\gamma z} e^{j\omega t} = -\omega^2 \mu\varepsilon E_{x0} e^{-\gamma z} e^{j\omega t}$$

$$so: \gamma^2 = -\omega^2 \mu\varepsilon, \quad or \quad \gamma = j\omega\sqrt{\mu\varepsilon}, \quad \pm\gamma \text{ are solutions}$$

- Which gives solution for propagation constant γ

Final Plane Wave Solution

$$\nabla^2 \mathbf{E} = -\mu\varepsilon \frac{\partial^2 \mathbf{E}}{\partial t^2}$$

$$\nabla^2 \mathbf{E} = \left(\frac{\partial^2}{\partial x^2} + \frac{\partial^2}{\partial y^2} + \frac{\partial^2}{\partial z^2} \right) \mathbf{E} = -\mu\varepsilon \frac{\partial^2 \mathbf{E}}{\partial t^2}$$

- Finally, the plane-wave solution from previous slide is:

$$with: \gamma^2 = -\omega^2 \mu\varepsilon, \quad or \quad \gamma = \pm j\omega\sqrt{\mu\varepsilon}$$

$$\mathbf{E} = \begin{bmatrix} E_{x0} e^{-\gamma z} e^{j\omega t} \\ 0 \\ 0 \end{bmatrix} = E_{x0} e^{-\gamma z} e^{j\omega t} \begin{bmatrix} 1 \\ 0 \\ 0 \end{bmatrix} = E_{x0} e^{\pm j\omega\sqrt{\mu\varepsilon}\, z} e^{j\omega t} \begin{bmatrix} 1 \\ 0 \\ 0 \end{bmatrix}$$

- And we will adopt a phasor shorthand "$\mathbf{E}_x e^{-\gamma z}$" defined as

$$definition: \mathbf{E}_x \, e^{-\gamma z} \Rightarrow \begin{bmatrix} E_{x0} e^{j\omega t} \\ 0 \\ 0 \end{bmatrix} e^{-\gamma z}, \, or \, \mathbf{E} \, e^{-\gamma z} \Rightarrow \begin{bmatrix} E_{x0} e^{j\omega t} \\ E_{y0} e^{j\omega t} \\ E_{z0} e^{j\omega t} \end{bmatrix} e^{-\gamma z}$$

- Beware: different authors use different phasor notations, such as Hayt's $\mathbf{E}_{sx}$, and Ellingson $\tilde{\mathbf{E}}_x$, and Ida $E_x(z)$

Propagation Constant γ and Wavelength: λ

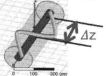

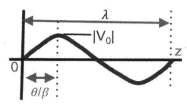

$\gamma = \alpha + j\beta$

$\lambda = 2\pi/\beta = v_p/f$

- For a traveling wave, wavelength λ is the physical length of one cycle in the transmission line, at an instant of time
- For a wave equation solution $\{|V_0| e^{j\theta} e^{j\omega t} e^{-\alpha z} e^{-j\beta z}\}$, the real wave would be $\text{Re}\{|V_0| e^{j(\omega t+\theta)} e^{-\alpha z} e^{-j\beta z}\} = |V_0| e^{-\alpha z} \cos(\omega t+\theta-\beta z)$
- For $\alpha=0$, a "snapshot" at t=0 would look like the figure above
- Wavelength λ is the physical distance x that completes one cycle, corresponding to $\beta z=2\pi$, or $\beta\lambda=2\pi$
- Thus, wavelength in meters: $\lambda = 2\pi/\beta = 2\pi/(\omega/v_p) = v_p/f$
- Since propagation constant $\gamma=\alpha+j\beta$, the imaginary part of γ determines the wavelength

Frequency and Spatial Frequency

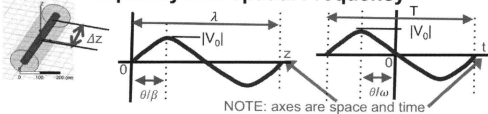

NOTE: axes are space and time

- For a traveling wave, in addition to the time-domain frequency ω in rad/s or f in Hz, there is a spatial frequency in rad/m
- For wave equation solution $\{|V_0| e^{j\theta} e^{j\omega t} e^{-\alpha z} e^{-j\beta z}\}$, the real wave is $\text{Re}\{|V_0| e^{j(\omega t+\theta)} e^{-\alpha z} e^{-j\beta z}\} = \cos(\omega t+\theta-\beta z)$ for $V_0=1$, $\alpha=0$
- At any point z_0 on the line, we would observe a time-varying signal $\cos(\omega t+\theta-\beta z_0)$ with frequency $\omega=2\pi f=2\pi/T$ rad/s or f Hz
- At any time instant t_0, we would observe a voltage along the line $\cos(\omega t_0+\theta-\beta z)$ with spatial frequency $\beta=2\pi/\lambda$ rad/m
- This spatial frequency is commonly called wavenumber and $k=\beta$ in textbooks using e^{-jkz} instead of $e^{-\gamma z}$ for lossless lines

Plane Wave Solution, t=0

$$\nabla^2 \mathbf{E} = -\mu\varepsilon \frac{\partial^2 \mathbf{E}}{\partial t^2}$$

- To visualize the plane-wave, "snapshot" wave at time t=0:

$$\mathbf{E_x}\, e^{-\gamma z} = \left[E_{x0}e^{j\omega t}\ 0\ 0 \right]^T e^{-\gamma z} = \left[E_{x0}e^{-\gamma z}\ 0\ 0 \right]^T @t=0$$

- Plotting the real part of the solution Re$\{E_{x0}e^{-\gamma z}\}$=E_{x0}cos($-\gamma$z)

Illustration of wave "snapshot" at t=0

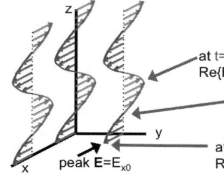

For simplicity, let $\mathbf{E_{x0}}$ be real

at t=0, and $\gamma z=j\pi$, E-field is
Re$\{E_{x0}e^{-\gamma z}\}$=Re$\{E_{x0}e^{-j\pi}\}$=E_{x0}cos($-\pi$)= $-E_{x0}$

at t=0, and $\gamma z=j\pi/2$, E-field is
Re$\{E_{x0}e^{-\gamma z}\}$=Re$\{E_{x0}e^{-j\pi/2}\}$= 0

at t=0, and z=0, E-field is peak
Re$\{E_{x0}e^{-\gamma z}\}$=E_{x0}cos(0)=E_{x0}

peak E=E_{x0}

- E-field has same value everywhere in x-y plane for a fixed z

Plane Wave Solution, t=π/(2ω)

$$\nabla^2 \mathbf{E} = -\mu\varepsilon \frac{\partial^2 \mathbf{E}}{\partial t^2}$$

- To visualize the plane-wave, "snapshot" wave at time t=π/(2ω):

$$\mathbf{E_x}\, e^{-\gamma z} = \left[E_{x0}e^{j\omega t}\ 0\ 0 \right]^T e^{-\gamma z} = \left[E_{x0}e^{-\gamma z}e^{j\pi/2}\ 0\ 0 \right]^T @t=\pi/(2\omega)$$

- Plotting the real part of solution Re$\{E_{x0}e^{-\gamma z}e^{j\pi/2}\}$=$E_{x0}$cos($\pi/2-\gamma$z)

Illustration of wave "snapshot" at t=π/(2ω)

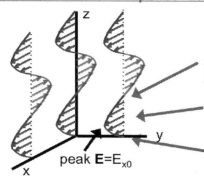

For simplicity, let $\mathbf{E_{x0}}$ be real

at t=π/(2ω), and $\gamma z=j\pi$, E-field is
Re$\{E_{x0}e^{-\gamma z}e^{j\pi/2}\}$=Re$\{E_{x0}e^{-j\pi}e^{j\pi/2}\}$= 0

at t=π/(2ω), and $\gamma z=j\pi/2$, E-field is peak
Re$\{E_{x0}e^{-\gamma z}e^{j\pi/2}\}$=Re$\{E_{x0}e^{-j\pi/2}e^{j\pi/2}\}$=$E_{x0}$

at t=π/(2ω), and z=0, E-field is
Re$\{E_{x0}e^{-\gamma z}e^{j\pi/2}\}$=$E_{x0}$cos($\pi/2$)=0

peak E=E_{x0}

- E-field has same value everywhere in x-y plane for a fixed z

Plane Wave Solution, t=0 and t=π/(2ω)

$$\nabla^2 \mathbf{E} = -\mu\varepsilon\frac{\partial^2 \mathbf{E}}{\partial t^2}$$

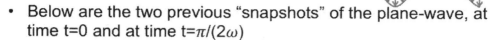

- Below are the two previous "snapshots" of the plane-wave, at time t=0 and at time t=π/(2ω)
- The plot at t=π/(2ω) appears to have moved upward by π/2 rad
- This is due to phase velocity v_p, as for the RC-line

Illustration of wave "snapshot" at t=0

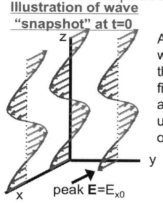

peak **E**=E_{x0}

At z=0, compared to the wave on the left, notice that the peak of the E-field on the right appears to have moved upward by a phase shift of π/2 radians

Illustration of wave "snapshot" at t=π/(2ω)

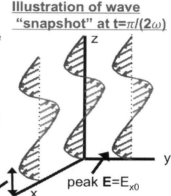

peak **E**=E_{x0}

π/2 rad

Plane Wave Solution, Phase Velocity

$$\nabla^2 \mathbf{E} = -\mu\varepsilon\frac{\partial^2 \mathbf{E}}{\partial t^2}; \quad \text{plane wave: } \mathbf{E_x}\, e^{-\gamma z} \Rightarrow \left[E_{x0}e^{j\omega t}\ 0\ 0 \right]^T e^{-\gamma z}$$

- The previous slide showed that the phase peaks of the plane wave were moving upward on the z-axis as time progressed
- This is due to phase velocity v_p, as before for the RC-line
- To find the phase velocity, recall that the RC-line wave-equation solution could be expressed as "A $e^{-\gamma x}$ $e^{j\omega t}$" or "A $e^{-\alpha x}$ $e^{-j\beta x}$ $e^{j\omega t}$" where $\beta = \omega/v_p$ where v_p was phase velocity
- By comparison, for the plane wave:

$$\mathbf{E_x}\, e^{-\gamma z} \Rightarrow \left[E_{x0}e^{-\gamma z}e^{j\omega t}\ 0\ 0 \right]^T, \gamma = \pm j\omega\sqrt{\mu\varepsilon} = \alpha + j\beta = \alpha + j\omega/v_p$$

$$\Rightarrow v_p = \frac{\omega}{\beta} = \frac{\omega}{\omega\,\mathrm{Re}\{\sqrt{\mu\varepsilon}\}} = \frac{1}{\mathrm{Re}\{\sqrt{\mu\varepsilon}\}}, \quad \boxed{\text{in vacuum, } v_p = \frac{1}{\sqrt{\mu_0\varepsilon_0}} = c = 3\times10^8}$$

- Thus, the phase velocity v_p of plane waves in vacuum is the speed of light ...for radio waves and light

Plane Wave Solution, Group Velocity

$$\nabla^2 \mathbf{E} = -\mu\varepsilon \frac{\partial^2 \mathbf{E}}{\partial t^2}; \quad \text{plane wave:} \ \mathbf{E_x}\, e^{-\gamma z} \Rightarrow \begin{bmatrix} E_{x0}e^{j\omega t} & 0 & 0 \end{bmatrix}^T e^{-\gamma z}$$

- Recall that group velocity is the speed of energy travel and was defined as $v_g = d\omega/d\beta$
- Using the result of the previous slide for the plane wave:

$$\mathbf{E_x}\, e^{-\gamma z} \Rightarrow \begin{bmatrix} E_{x0}e^{-\gamma z}e^{j\omega t} & 0 & 0 \end{bmatrix}^T, \gamma = \pm j\omega\sqrt{\mu\varepsilon} = \alpha + j\beta = \alpha + j\omega/v_p$$

$$\Rightarrow v_g = \frac{d\omega}{d\beta} = \left(\frac{d\beta}{d\omega}\right)^{-1} = \left(\frac{d}{d\omega}\omega\,\mathrm{Re}\{\sqrt{\mu\varepsilon}\}\right)^{-1} = \frac{1}{\mathrm{Re}\{\sqrt{\mu\varepsilon}\}} = v_p$$

$$\boxed{\text{in vacuum, } v_g = v_p = \frac{1}{\sqrt{\mu_0\varepsilon_0}} = c = 3\times 10^8}$$

- Thus, phase velocity v_p and group velocity v_g of plane waves in vacuum are the speed of light ...for radio waves and light

Plane Wave H-Field Solution

$$\nabla^2 \mathbf{E} = -\mu\varepsilon \frac{\partial^2 \mathbf{E}}{\partial t^2}; \quad \text{plane wave:} \ \mathbf{E_x}\, e^{-\gamma z} \Rightarrow \begin{bmatrix} E_{x0}e^{j\omega t} & 0 & 0 \end{bmatrix}^T e^{-\gamma z}$$

- To find the solution for the H-field, substitute the E-field solution into Faraday's law: $\nabla \times \mathbf{E} = -\frac{\partial \mathbf{B}}{\partial t} = -\mu\frac{\partial \mathbf{H}}{\partial t}$

- Substituting:

$$-\mu\frac{\partial \mathbf{H}}{\partial t} = \nabla \times \mathbf{E} = \nabla \times \begin{bmatrix} E_{x0}e^{j\omega t} \\ 0 \\ 0 \end{bmatrix}e^{-\gamma z} = \begin{vmatrix} \hat{\mathbf{x}} & \hat{\mathbf{y}} & \hat{\mathbf{z}} \\ \partial/\partial x & \partial/\partial y & \partial/\partial z \\ E_{x0}e^{-\gamma z}e^{j\omega t} & 0 & 0 \end{vmatrix}$$

$$= \left(\frac{\partial}{\partial y}0 - \frac{\partial}{\partial z}0\right)\hat{\mathbf{x}} - \left(\frac{\partial}{\partial x}0 - \frac{\partial}{\partial z}E_{x0}e^{-\gamma z}e^{j\omega t}\right)\hat{\mathbf{y}} + \left(\frac{\partial}{\partial x}0 - \frac{\partial}{\partial y}E_{x0}e^{-\gamma z}e^{j\omega t}\right)\hat{\mathbf{z}}$$

d/dy term =0

$$-\mu\frac{\partial \mathbf{H}}{\partial t} = \frac{\partial}{\partial z}E_{x0}e^{-\gamma z}e^{j\omega t}\,\hat{\mathbf{y}} = -\gamma E_{x0}e^{-\gamma z}e^{j\omega t}\,\hat{\mathbf{y}} \quad \Rightarrow j\omega\mu\mathbf{H} = \gamma E_{x0}e^{-\gamma z}e^{j\omega t}\,\hat{\mathbf{y}}$$

$$\boxed{\mathbf{H} = \frac{\gamma}{j\omega\mu}E_{x0}e^{-\gamma z}e^{j\omega t}\,\hat{\mathbf{y}} = \frac{\gamma}{j\omega\mu}\begin{bmatrix} 0 \\ E_{x0}e^{-\gamma z}e^{j\omega t} \\ 0 \end{bmatrix} \quad \substack{\text{From the E-field solution}} \quad where: \gamma = \pm j\omega\sqrt{\mu\varepsilon}}$$

Plane Wave H Solution and Intrinsic Impedance η

$$\nabla^2 \mathbf{E} = -\mu\varepsilon \frac{\partial^2 \mathbf{E}}{\partial t^2}; \quad \text{plane wave: } \mathbf{E_x}\, e^{-\gamma z} \Rightarrow \left[E_{x0}e^{j\omega t}\ 0\ 0 \right]^T e^{-\gamma z}$$

- Just as transmission-line impedance Z_0 determines the ratio of voltage to current in the RC-line, intrinsic impedance η determines the ration of E-field to H-field for the plane wave

- From the previous slide, the plane-wave solution for the H-field in vector form is

$$\text{for } \mathbf{E} = \begin{bmatrix} E_{x0}e^{-\gamma z}e^{j\omega t} \\ 0 \\ 0 \end{bmatrix} \quad \text{then } \mathbf{H} = \frac{\gamma}{j\omega\mu}\begin{bmatrix} 0 \\ E_{x0}e^{-\gamma z}e^{j\omega t} \\ 0 \end{bmatrix} = \frac{1}{\eta}\begin{bmatrix} 0 \\ E_{x0}e^{-\gamma z}e^{j\omega t} \\ 0 \end{bmatrix}$$

or: $\mathbf{H} = \dfrac{\gamma}{j\omega\mu} E_{x0}e^{-\gamma z} e^{j\omega t}\ \hat{\mathbf{y}} = \dfrac{1}{\eta} E_{x0}e^{-\gamma z} e^{j\omega t}\ \hat{\mathbf{y}} \quad where: \gamma = j\omega\sqrt{\mu\varepsilon}$

$$\text{intrinsic impedance } \eta = \frac{j\omega\mu}{\gamma} = \frac{j\omega\mu}{j\omega\sqrt{\mu\varepsilon}} = \sqrt{\frac{\mu}{\varepsilon}} \quad and\ \eta_0 = \sqrt{\frac{\mu_0}{\varepsilon_0}} = 377\Omega$$

in vacuum

Plane Wave E-field and H-Field Phasors

$$\nabla^2 \mathbf{E} = -\mu\varepsilon \frac{\partial^2 \mathbf{E}}{\partial t^2}; \quad \text{plane wave: } \mathbf{E_x}\, e^{-\gamma z} \Rightarrow \left[E_{x0}e^{j\omega t}\ 0\ 0 \right]^T e^{-\gamma z}$$

- The phasor form for the plane-wave solutions is then

$$\mathbf{E} \text{ definition: } \mathbf{E_x}e^{-\gamma z} \Rightarrow \begin{bmatrix} E_{x0}e^{j\omega t} \\ 0 \\ 0 \end{bmatrix}e^{-\gamma z} = \begin{bmatrix} E_{x0}e^{-\gamma z}e^{j\omega t} \\ 0 \\ 0 \end{bmatrix} = \begin{bmatrix} E_x \\ 0 \\ 0 \end{bmatrix}$$

$$\text{so, } \mathbf{H} \text{ phasor definition: } \mathbf{H_y}e^{-\gamma z} \Rightarrow \begin{bmatrix} 0 \\ H_{y0}e^{j\omega t} \\ 0 \end{bmatrix}e^{-\gamma z} = \frac{1}{\eta}\begin{bmatrix} 0 \\ E_{x0}e^{j\omega t} \\ 0 \end{bmatrix}e^{-\gamma z}$$

$$\text{where intrinsic impedance } \eta = \frac{j\omega\mu}{\gamma} = \sqrt{\frac{\mu}{\varepsilon}} = \frac{E_{x0}}{H_{y0}} \quad where: \gamma = \pm j\omega\sqrt{\mu\varepsilon}$$

- So, $\eta = E_{x0}/H_{y0}$, the ratio of the E-field to H-field components

Plane Wave Solution E-field and H-field, $t=\pi/(2\omega)$

- Visualize the plane-wave, "snapshot" wave at time $t=\pi/(2\omega)$:

$$\mathbf{E_x}\, e^{-\gamma z}=\left[E_{x0}e^{j\omega t}\ 0\ 0\right]^{T} e^{-\gamma z} =\left[E_{x0}e^{-\gamma z}e^{j\pi/2}\ 0\ 0\right]^{T} @t=\pi/(2\omega)$$

$$\mathbf{H_y}\, e^{-\gamma z}=\left[0\ H_{y0}e^{j\omega t}\ 0\right]^{T} e^{-\gamma z} =\left[0\ H_{y0}e^{-\gamma z}e^{j\pi/2}\ 0\right]^{T} @t=\pi/(2\omega)$$

- Plot real part of solution $\mathrm{Re}\{E_{x0}e^{-\gamma z}e^{j\pi/2}\}$ and $\mathrm{Re}\{H_{y0}e^{-\gamma z}e^{j\pi/2}\}$

Illustration of wave "snapshot" at $t=\pi/(2\omega)$

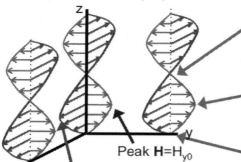

Peak H=H_{y0}

peak E=E_{x0}

For simplicity, let E_{x0} snd H_{y0} be real

at $t=\pi/(2\omega)$, and $\gamma z=j\pi$, E-field is
$\mathrm{Re}\{E_{x0}e^{-\gamma z}e^{j\pi/2}\}=\mathrm{Re}\{E_{x0}e^{-j\pi}e^{j\pi/2}\}=0$
And H-field is $\mathrm{Re}\{H_{y0}e^{-\gamma z}e^{j\pi/2}\}=0$

at $t=\pi/(2\omega)$, and $\gamma z=j\pi/2$, E-field is peak
$\mathrm{Re}\{E_{x0}e^{-\gamma z}e^{j\pi/2}\}=\mathrm{Re}\{E_{x0}e^{-j\pi/2}e^{j\pi/2}\}=E_{x0}$
And H-field is peak $\mathrm{Re}\{H_{y0}e^{-j\pi/2}e^{j\pi/2}\}=H_{y0}$

at $t=\pi/(2\omega)$, and $z=0$, E-field is
$\mathrm{Re}\{E_{x0}e^{-\gamma z}e^{j\pi/2}\}=E_{x0}\cos(\pi/2)=0$
And H-field is $\mathrm{Re}\{H_{y0}e^{-\gamma z}e^{j\pi/2}\}=0$

Phasor Form of Wave Equation and Solution

$$\nabla^2\mathbf{E} = -\mu\varepsilon\frac{\partial^2\mathbf{E}}{\partial t^2};\ \text{ plane wave: }\ \mathbf{E_x}\, e^{-\gamma z} \Rightarrow \left[E_{x0}e^{j\omega t}\ 0\ 0\right]^{T} e^{-\gamma z}$$

- The phasor form for the wave equation is similarly derived:

$$\nabla\times\mathbf{E} = -j\omega\,\mathbf{B} \qquad\qquad \nabla\times\mathbf{H} = \mathbf{J_e} + j\omega\,\mathbf{D} = j\omega\,\mathbf{D}\ \text{ for }\mathbf{J_e} = 0$$

curl of Faraday $\Rightarrow \nabla\times\nabla\times\mathbf{E} = -\nabla\times j\omega\,\mathbf{B}\ \Rightarrow \nabla(\nabla\cdot\mathbf{E}) - \nabla^2\mathbf{E} = -j\omega\mu\nabla\times\mathbf{H}$

then in source-free region $\nabla^2\mathbf{E} = j\omega\mu\nabla\times\mathbf{H}$ because $(\nabla\cdot\mathbf{E}) = 0$

then substitute ampere $\Rightarrow \nabla^2\mathbf{E} = j\omega\mu(j\omega\,\mathbf{D}) = -\omega^2\mu\mathbf{D} = -\omega^2\mu\varepsilon\mathbf{E}$

$\Rightarrow$ phasor form of wave equation: $\nabla^2\mathbf{E} = -\omega^2\mu\varepsilon\mathbf{E}$

- And substitute generic phasor solution $\mathbf{E_x}e^{-\gamma z}$, to get solution:

$$\nabla^2\mathbf{E} = -\omega^2\mu\varepsilon\mathbf{E} \Rightarrow \nabla^2\mathbf{E_x}e^{-\gamma z} = \left(\frac{\partial^2}{\partial x^2}+\frac{\partial^2}{\partial y^2}+\frac{\partial^2}{\partial z^2}\right)\mathbf{E_x}e^{-\gamma z} = -\omega^2\mu\varepsilon\mathbf{E_x}e^{-\gamma z}$$

$$\Rightarrow \gamma^2\mathbf{E_x}e^{-\gamma z} = -\omega^2\mu\varepsilon\mathbf{E_x}e^{-\gamma z},\ \text{ and so }\ \gamma^2 = -\omega^2\mu\varepsilon,\ or\ \gamma = \pm j\omega\sqrt{\mu\varepsilon}$$

- Thus, phasor form gives the same E-field as before

Plane Wave Phasor H-Field Solution

$$\nabla^2 \mathbf{E} = -\mu\varepsilon \frac{\partial^2 \mathbf{E}}{\partial t^2}; \quad \text{plane wave: } \mathbf{E_x}\, e^{-\gamma z} \Rightarrow \left[E_{x0} e^{j\omega t}\ 0\ 0 \right]^T e^{-\gamma z}$$

- Similarly, the phasor solution for the H-field can be found by substituting the phasor E-field solution into Faraday's law:

$$-j\omega\, \mathbf{B} = \nabla \times \mathbf{E} = \nabla \times \mathbf{E_x} e^{-\gamma z} = \nabla \times \begin{bmatrix} E_{x0} e^{-\gamma z} \\ 0 \\ 0 \end{bmatrix} = \begin{vmatrix} \hat{\mathbf{x}} & \hat{\mathbf{y}} & \hat{\mathbf{z}} \\ \partial/\partial x & \partial/\partial y & \partial/\partial z \\ E_{x0} e^{-\gamma z} & 0 & 0 \end{vmatrix}$$

$$= \left(\frac{\partial}{\partial y} 0 - \frac{\partial}{\partial z} 0 \right)\hat{\mathbf{x}} - \left(\frac{\partial}{\partial x} 0 - \frac{\partial}{\partial z} E_{x0} e^{-\gamma z} \right)\hat{\mathbf{y}} + \left(\frac{\partial}{\partial x} 0 - \frac{\partial}{\partial y} E_{x0} e^{-\gamma z} \right)\hat{\mathbf{z}}$$

d/dy term =0

$$\Rightarrow -j\omega\, \mathbf{B} = -j\omega\mu\, \mathbf{H} = \frac{\partial}{\partial z} E_{x0} e^{-\gamma z}\, \hat{\mathbf{y}} = -\gamma E_{x0} e^{-\gamma z}\, \hat{\mathbf{y}}$$

$$\Rightarrow \mathbf{H} = \frac{\gamma}{j\omega\mu} E_{x0} e^{-\gamma z}\, \hat{\mathbf{y}} = \frac{\gamma}{j\omega\mu} \begin{bmatrix} 0 \\ E_{x0} e^{-\gamma z} \\ 0 \end{bmatrix} = \frac{1}{\eta} \begin{bmatrix} 0 \\ E_{x0} e^{-\gamma z} \\ 0 \end{bmatrix}$$

- Thus, phasor form gives the same H-field as before

Phasor Form Solution: Plane Wave in Lossy Dielectric

$$\nabla^2 \mathbf{E} = -\mu\varepsilon \frac{\partial^2 \mathbf{E}}{\partial t^2}; \quad \text{plane wave: } \mathbf{E_x}\, e^{-\gamma z} \Rightarrow \left[E_{x0} e^{j\omega t}\ 0\ 0 \right]^T e^{-\gamma z}$$

- Now the wave equation is derived for lossy dielectric with $\mathbf{J_e} \neq 0$:

$$\nabla \times \mathbf{E} = -j\omega\, \mathbf{B} \qquad \nabla \times \mathbf{H} = \mathbf{J_e} + j\omega\, \mathbf{D} = (\sigma + j\omega\varepsilon)\mathbf{E} \quad \text{for } \mathbf{J_e} \neq 0$$

curl of Faraday $\Rightarrow \nabla \times \nabla \times \mathbf{E} = -\nabla \times j\omega\, \mathbf{B} \Rightarrow \nabla(\nabla \cdot \mathbf{E}) - \nabla^2 \mathbf{E} = -j\omega\mu\nabla \times \mathbf{H}$

then in source-free region $\nabla^2 \mathbf{E} = j\omega\mu\nabla \times \mathbf{H}$ because $(\nabla \cdot \mathbf{E}) = 0$

substitute ampere $\Rightarrow \nabla^2 \mathbf{E} = j\omega\mu(\sigma + j\omega\varepsilon)\mathbf{E} = -\omega^2\mu\varepsilon(1 - j\sigma/(\omega\varepsilon))\mathbf{E}$

$\Rightarrow$ phasor form of wave equation: $\nabla^2 \mathbf{E} = -\omega^2\mu\varepsilon(1 - j\sigma/(\omega\varepsilon))\mathbf{E}$

- And substitute "generic" phasor solution $\mathbf{E_x} e^{-\gamma z}$, as before:

$$\nabla^2 \mathbf{E} = -\omega^2\mu\varepsilon(1 - j\sigma/(\omega\varepsilon))\mathbf{E}$$

Note: $\gamma = \alpha + j\beta$ where α is no longer zero

$$\Rightarrow \nabla^2 \mathbf{E_x} e^{-\gamma z} = \gamma^2 \mathbf{E_x} e^{-\gamma z} = -\omega^2\mu\varepsilon(1 - j\sigma/(\omega\varepsilon))\mathbf{E_x} e^{-\gamma z}$$

and so $\gamma^2 = -\omega^2\mu\varepsilon(1 - j\sigma/(\omega\varepsilon))$, or $\gamma = \pm j\omega\sqrt{\mu\varepsilon}\sqrt{1 - j\sigma/(\omega\varepsilon)}$

similarly: $\mathbf{H} = \dfrac{1}{\eta} E_{x0} e^{-\gamma z}\, \hat{\mathbf{y}} = \dfrac{\gamma}{j\omega\mu}$ where: $\eta = \dfrac{j\omega\mu}{\gamma}$

Field Energy and Poynting Vector

Energy In Electromagnetic Field

- To find the energy in the electromagnetic field

with: $\nabla \times \mathbf{E} = -j\omega\, \mathbf{B}$ and $\nabla \times \mathbf{H} = \mathbf{J_e} + \partial \mathbf{D}/\partial t$

use vector identity: $\mathbf{A} \cdot (\nabla \times \mathbf{B}) - \mathbf{B} \cdot (\nabla \times \mathbf{A}) = \nabla \cdot (\mathbf{B} \times \mathbf{A})$

$\Rightarrow \mathbf{H} \cdot (\nabla \times \mathbf{E}) - \mathbf{E} \cdot (\nabla \times \mathbf{H}) = \nabla \cdot (\mathbf{E} \times \mathbf{H})$

substitute: $\nabla \times \mathbf{E} = -\partial \mathbf{B}/\partial t$ and $\nabla \times \mathbf{H} = \mathbf{J_e} + \partial \mathbf{D}/\partial t$

$\Rightarrow -\mathbf{H} \cdot \partial \mathbf{B}/\partial t - \mathbf{E} \cdot (\mathbf{J_e} + \partial \mathbf{D}/\partial t) = \nabla \cdot (\mathbf{E} \times \mathbf{H})$

volume integral: $\int_V (\mathbf{H} \cdot \partial \mathbf{B}/\partial t + \mathbf{E} \cdot \partial \mathbf{D}/\partial t + \mathbf{E} \cdot \mathbf{J_e})dv = -\int_V \nabla \cdot (\mathbf{E} \times \mathbf{H})dv$

apply divergence theorem to right side to obtain final result:

Poynting's Theorem: $\int_V (\mathbf{H} \cdot \partial \mathbf{B}/\partial t + \mathbf{E} \cdot \partial \mathbf{D}/\partial t + \mathbf{E} \cdot \mathbf{J_e})dv = -\oint_S (\mathbf{E} \times \mathbf{H}) \cdot d\mathbf{S}$

Change in stored magnetic energy

Change in stored electric energy

dissipated Resistive heat

Power into volume

- Showing balance of energy in volume with power crossing surface

Credit: Ramo, Whinnery, VanDuzer, "Fields and Waves in Comm. Electronics," JWiley&Sons, 1984.

100

Energy In Electromagnetic Field and Poynting Vector

- Before proceeding to the Poynting vector, note that for a linear, time-invariant, isotropic medium, the energy in the electromagnetic field can be rewritten

$$\int_V \left(\mathbf{H} \cdot \partial \mathbf{B}/\partial t + \mathbf{E} \cdot \partial \mathbf{D}/\partial t + \mathbf{E} \cdot \mathbf{J}_e\right) dv = \oint_S \left(\mathbf{E} \times \mathbf{H}\right) \cdot d\mathbf{S}$$

$$\Rightarrow \int_V \left(\frac{\partial}{\partial t}\left(\frac{\mathbf{B} \cdot \mathbf{H}}{2}\right) + \frac{\partial}{\partial t}\left(\frac{\mathbf{D} \cdot \mathbf{E}}{2}\right) + \mathbf{E} \cdot \mathbf{J}_e\right) dv = -\oint_S \left(\mathbf{E} \times \mathbf{H}\right) \cdot d\mathbf{S}$$

Change in stored magnetic energy   Change in stored electric energy   dissipated Resistive heat   Power into volume (because of minus sign)

- Note that the magnetic and electric energy had the same forms for the static fields
- Finally, the vector **E×H** above is the Poynting vector, and describes the direction and magnitude of power flux in W/m²:

$$\text{Poynting vector: } \mathbf{P} = \mathbf{E} \times \mathbf{H}$$

Credit: Ramo, Whinnery, VanDuzer, "Fields and Waves in Comm. Electronics," JWiley&Sons,1984.

Phasor Form of Field Energy and Poynting Vector

- Note that the preceding Poynting Theorem is in time domain, the phasor form is presented here without proof:

$$\int_V \left(j\omega \mathbf{H}^* \cdot \mathbf{B} - j\omega \mathbf{E} \cdot \mathbf{D}^* + \mathbf{E} \cdot \mathbf{J}_e^*\right) dv = -\oint_S \mathbf{E} \times \mathbf{H}^* \cdot d\mathbf{S}$$

$$\Rightarrow j2\omega \int_V \left(\frac{\mathbf{B} \cdot \mathbf{H}^*}{4} - \frac{\mathbf{D} \cdot \mathbf{E}^*}{4}\right) dv + \int_V \frac{\mathbf{E} \cdot \mathbf{J}_e^*}{2} dv = -\oint_S \frac{\mathbf{E} \times \mathbf{H}^*}{2} \cdot d\mathbf{S}$$

Reactive time-averaged power into volume

$$\Rightarrow j2\omega \int_V \left(\frac{\mathbf{B} \cdot \mathbf{H}^*}{4} - \frac{\mathbf{D} \cdot \mathbf{E}^*}{4}\right) dv + \int_V \frac{\mathbf{E} \cdot \mathbf{J}_e^*}{2} dv = -\oint_S \left(\text{Re}\left(\frac{\mathbf{E} \times \mathbf{H}^*}{2}\right) + j\,\text{Im}\left(\frac{\mathbf{E} \times \mathbf{H}^*}{2}\right)\right) \cdot d\mathbf{S}$$

Time averaged magnetic energy stored in volume   Time averaged electric energy stored in volume   Dissipated power inside volume   Real time-averaged power into volume

Beware minus sign here

- Note that the magnetic and electric energy are reactive power flows with "jω" outside the integral

Credit: Ramo, Whinnery, VanDuzer, "Fields and Waves in Comm. Electronics," JWiley&Sons,1984.

Real and Reactive Power for Phasors

- The complex Poynting vector (or phasor Poynting vector) has a form where $\mathbf{E}\times\mathbf{H}^*$, and the time average real power is defined as $P_{avg\_real}=0.5\ \text{Re}\{\mathbf{E}\times\mathbf{H}^*\}$

- To see why, consider the time-domain version of of phasor voltage source $\mathbf{V}=ve^{j\theta}$ driving a reactive load of an inductor in series with a capacitor with $\mathbf{Z}=R+jX_L+jX_C=ze^{j\varphi}$. The phasor current is then $\mathbf{I}=\mathbf{V}/\mathbf{Z}=(v/z)e^{j(\theta-\varphi)}$. Real average power is only delivered to the resistor, and since the current $\mathbf{I}$ must flow through R and is a sinusoid of amplitude v/z, the rms time-averaged power delivered to R is $P_{avg\_real}=(v/z)^2R/2$. By comparison, we see that the phasor product $\mathbf{VI}^*=ve^{j\theta}(v/z)e^{-j(\theta-\varphi)}=v(v/z)e^{j\varphi}=(v/z)^2ze^{j\varphi}=(v/z)^2\mathbf{Z}$ $=(v/z)^2(R+jX_L+jX_C)$. Finally, it is clear that $\text{Re}\{\mathbf{VI}^*\}=(v/z)^2R=2P_{avg\_real}$. Thus, $P_{avg\_real}=0.5\text{Re}\{\mathbf{VI}^*\}$ corresponds to $P_{avg\_real}=0.5\ \text{Re}\{\mathbf{E}\times\mathbf{H}^*\}$.

- Also, note that stored energy in an inductor is $0.5\ L\ i^2$, and for a sinusoidal peak current $|\mathbf{I}|$, the time-average energy is $E_L=L\ |\mathbf{I}|^2/4$. The stored energy in an capacitor is $0.5\ C\ v^2$, and for a sinusoidal peak current $|\mathbf{V}|$, the time-average energy is $E_C=C\ |\mathbf{V}_C|^2/4$. Comparing the forgoing phasor product, $\text{Im}\{\mathbf{VI}^*\}=(v/z)^2(X_L+X_C)=|\mathbf{I}|^2\omega L-|\mathbf{I}|^2/(\omega C)=|\mathbf{I}|^2\omega L-|\mathbf{V}_C\omega C|^2/(\omega C)=|\mathbf{I}|^2\omega L-|\mathbf{V}_C|^2\omega C$, so $\text{Im}\{\mathbf{VI}^*\}=4\omega(E_L-E_C)$. Thus, $0.5\text{Im}\{\mathbf{VI}^*\}=2\omega(E_L-E_C)$ corresponds to the complex Poynting imaginary component $0.5\text{Im}\{\mathbf{E}\times\mathbf{H}^*\}$.

Credit: Ramo, Whinnery, VanDuzer, "Fields and Waves in Comm. Electronics," JWiley&Sons,1984.

Complex (Phasor) Poynting Vector of Plane Wave

- Consider the complex Poynting vector of plane-wave solution

$$\mathbf{E_x}\ e^{-\gamma z}\Rightarrow\begin{bmatrix}E_{x0}e^{j\omega t}&0&0\end{bmatrix}^T e^{-\gamma z}\ ,\quad \mathbf{H_y}\ e^{-\gamma z}\Rightarrow\begin{bmatrix}0&(1/\eta)E_{x0}e^{j\omega t}&0\end{bmatrix}^T e^{-\gamma z}$$

complex Poynting vector is then:

$$\mathbf{E}\times\mathbf{H}^*=\left(\mathbf{E_x}\ e^{-\gamma z}\right)\times\left(\mathbf{H_y}\ e^{-\gamma z}\right)^*$$

$$=\begin{vmatrix}\hat{\mathbf{x}}&\hat{\mathbf{y}}&\hat{\mathbf{z}}\\ E_{x0}e^{-\gamma z}&0&0\\ 0&\left((1/\eta)E_{x0}e^{-\gamma z}\right)^*&0\end{vmatrix}$$

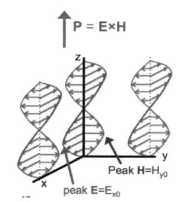

$$=(1/\eta)^*\left|E_{x0}e^{-\gamma z}\right|^2\hat{\mathbf{z}}$$

- Note that $\mathbf{E}\times\mathbf{H}^*$ follows right-hand rule
- For the earlier plane wave example, direction of $\mathbf{P}$ is illustrated, and is in same direction as group velocity

Example: Complex Poynting Vector of Plane Wave

- Find complex Poynting vector and average power density for a plane wave in vacuum with 1 V/m E-field, oriented in x direction

$$\mathbf{E_x}\, e^{-\gamma z} \Rightarrow \begin{bmatrix} E_{x0} e^{j\omega t} & 0 & 0 \end{bmatrix}^T e^{-\gamma z}\,, \quad \mathbf{H_y}\, e^{-\gamma z} \Rightarrow \begin{bmatrix} 0 & (1/\eta) E_{x0} e^{j\omega t} & 0 \end{bmatrix}^T e^{-\gamma z}$$

$$\gamma = \alpha + j\beta = j\omega\sqrt{\mu_0 \varepsilon_0} = j\omega / c$$

$$\mathbf{E}\times\mathbf{H}^* = \left(\mathbf{E_x}\, e^{-\gamma z}\right) \times \left(\mathbf{H_y}\, e^{-\gamma z}\right)^*$$

$$= \begin{vmatrix} \hat{\mathbf{x}} & \hat{\mathbf{y}} & \hat{\mathbf{z}} \\ (1)e^{-j\omega z/c} & 0 & 0 \\ 0 & \left((1/377)(1)e^{-j\omega z/c}\right)^* & 0 \end{vmatrix}$$

$$= (1/377)^* \left|1 e^{-j\omega z/c}\right|^2 \hat{\mathbf{z}} = (1/377)\hat{\mathbf{z}}$$

$$P_{avg-real} = (1/2)\,\mathrm{Re}\{\mathbf{E}\times\mathbf{H}^*\} = (1/754)\hat{\mathbf{z}}\ \ \mathrm{W/m}^2$$

P = E×H

Peak H=H_{y0}

peak E=E_{x0}

Conductors and Skin Depth

Skin Depth

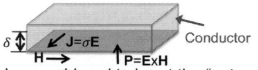

- Conductors can be considered to be at the "extreme" of the lossy plane-wave solution, with $\sigma \gg \omega\varepsilon$, then:

$$\mathbf{E_x}\, e^{-\gamma z} \Rightarrow \left[E_{x0}e^{j\omega t}\; 0\; 0 \right]^T e^{-\gamma z}\,, \quad \mathbf{H_y}\, e^{-\gamma z} \Rightarrow \left[0\; (1/\eta) E_{x0}e^{j\omega t}\; 0 \right]^T e^{-\gamma z}$$

propagation constant: $\gamma = \alpha + j\beta = j\omega\sqrt{\mu\varepsilon}\sqrt{1 - j\sigma/(\omega\varepsilon)}$

for $\sigma \gg \omega\varepsilon \;\Rightarrow\; \gamma \approx j\omega\sqrt{\mu\varepsilon}\sqrt{-j\sigma/(\omega\varepsilon)} = \sqrt{j\omega\mu\sigma} = \sqrt{\omega\mu\sigma/2}\,(1+j)$

then : $\mathbf{E_x}\, e^{-\gamma z} = \mathbf{E_x}\, e^{-\sqrt{\omega\mu\sigma/2}(1+j)z} = \mathbf{E_x}\, e^{-\sqrt{\omega\mu\sigma/2}\, z}\; e^{-j\sqrt{\omega\mu\sigma/2}\, z}$

at $z = 0$, E-field is $\left| \mathbf{E_x}\, e^{-\sqrt{\omega\mu\sigma/2}z}\; e^{-j\sqrt{\omega\mu\sigma/2}z} \right| = \left| \mathbf{E_x}\, e^0 \right| = \left| \mathbf{E_x} \right|$

at $z = \sqrt{2/(\omega\mu\sigma)}$, E-field is $\left| \mathbf{E_x}\, e^{-\sqrt{\omega\mu\sigma/2}\, z}\; e^{-j\sqrt{\omega\mu\sigma/2}\, z} \right|$

$$= \left| \mathbf{E_x}\, e^{-1}\, e^{-j} \right| = e^{-1} \left| \mathbf{E_x} \right|$$

so, at a distance of $z = \sqrt{2/(\omega\mu\sigma)}$ the E-field decreases by e^{-1} or 8.7 dB

the distance $\delta_s = \sqrt{2/(\omega\mu\sigma)}$ is called the skin depth of a conductor

Example: Conductor Skin Depth

- Find the skin depth of a copper conductor at 60 Hz and at 1 GHz with $\varepsilon_r=1$, $\mu_r=1$, and $\sigma= 6\times10^7$ S/m,

- At 60 Hz:

first check for $\sigma \gg \omega\varepsilon$, $\quad \dfrac{\sigma}{\omega\varepsilon} = \dfrac{6\times10^7}{(2\pi60)(8.85\times10^{-12})} = 4.1\times10^{15}$

then, skin depth:

$$\delta_s = \sqrt{2/(\omega\mu\sigma)} = \sqrt{2/\{(2\pi60)(1257\times10^{-9})(6\times10^7)\}} = 8.4 \text{ mm}$$

- At 1 GHz:
first check for $\sigma \gg \omega\varepsilon$, $\quad \dfrac{\sigma}{\omega\varepsilon} = \dfrac{6\times10^7}{(2\pi10^9)(8.85\times10^{-12})} = 4.1\times10^{15}$

then, skin depth:

$$\delta_s = \sqrt{2/(\omega\mu\sigma)} = \sqrt{2/\{(2\pi10^9)(1257\times10^{-9})(6\times10^7)\}} = 2.1\ \mu\text{m}$$

Polarization

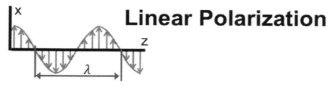

Linear Polarization

- Linear polarization is defined as a fixed orientation of the E-field as a plane wave travels
- As illustrated above, the E-field vector **E=E$_x$** only contains components along one axis (x-axis) even as it travels up the z-axis, and is referred to as linear polarization (linearly polarized)
- We could have just as easily defined a linearly polarized plane wave with E-field **E$_y$** containing only y-axis E-field components
- We also can use a vector sum to define a linearly polarized pane wave of any orientation in the x-y plane with:

$$\mathbf{E} = \left(\mathbf{E_x} + \mathbf{E_y} \right) e^{-\gamma z} \Rightarrow \left(\left[E_{x0}e^{j\omega t}\ 0\ 0 \right]^T + \left[0\ E_{y0}e^{j\omega t}\ 0 \right]^T \right) e^{-\gamma z}$$

- The sum of 2 orthogonal linear polarizations in the x-y plane can create any linear polarization in the x-y plane

Circular And Elliptical Polarization

- Circular polarization is defined as a rotating orientation of the E-field as a plane wave travels

- It is constructed by 2 equal-amplitude orthogonal polarizations in a plane, but 90 degree phase shifted in time, such as:

$$\mathbf{E} = \left(\mathbf{E_x} + \mathbf{E_y} e^{-j\pi/2} \right) e^{-\gamma z} \Rightarrow \left(\left[E_0 e^{j\omega t}\ 0\ 0 \right]^T + \left[0\ E_0 e^{j(\omega t - \pi/2)}\ 0 \right]^T \right) e^{-\gamma z}$$

- At any fixed plane on the z-axis, the E-field appears to rotate 360 degrees every time cycle of the frequency of the wave

- At any fixed time "snapshot" an E-field would appear to "corkscrew" in space along the z-axis

- Elliptical polarization uses 2 unequal amplitude orthogonal polarizations in a plane, 90 degree phase shifted in time:

$$\mathbf{E} = \left(\mathbf{E_x} + \mathbf{E_y} e^{-j\pi/2} \right) e^{-\gamma z} \Rightarrow \left(\left[E_{x0} e^{j\omega t}\ 0\ 0 \right]^T + \left[0\ E_{y0} e^{j(\omega t - \pi/2)}\ 0 \right]^T \right) e^{-\gamma z}$$

Left/Right Circular And Elliptical Polarization

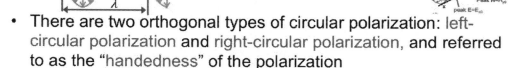

- There are two orthogonal types of circular polarization: left-circular polarization and right-circular polarization, and referred to as the "handedness" of the polarization

- Right-hand is defined by using the right hand with thumb pointed in direction of propagation (Poynting vector), where the fingers indicate the direction of rotation in time at a fixed point

- Right-polarized example:

$$\mathbf{E} = \left(\mathbf{E_x} + \mathbf{E_y} e^{-j\pi/2} \right) e^{-\gamma z} \Rightarrow \left(\left[E_0 e^{j\omega t}\ 0\ 0 \right]^T + \left[0\ E_0 e^{j(\omega t - \pi/2)}\ 0 \right]^T \right) e^{-\gamma z}$$

at z=0 and t=0: $\text{Re}\{\mathbf{E}\} = \text{Re}\left\{ \left[E_0 e^{j\omega t}\ E_0 e^{j(\omega t - \pi/2)}\ 0 \right]^T \right\} = \left[E_0\ 0\ 0 \right]^T$

at z=0, t=$\pi/(2\omega)$: $\text{Re}\{\mathbf{E}\} = \text{Re}\left\{ \left[E_0 e^{j\pi/2}\ E_0 e^{j(\pi/2 - \pi/2)}\ 0 \right]^T \right\} = \left[0\ E_0\ 0 \right]^T$

Summary

Plane Wave

Summary: Plane Wave Solution (Lossless)

wave equation: $\nabla^2 \mathbf{E} = \mu\varepsilon \dfrac{\partial^2 \mathbf{E}}{\partial t^2}$;

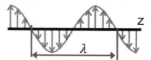

plane wave solution:

$$\mathbf{E_x}\, e^{-\gamma z} \Rightarrow \left[E_{x0} e^{j\omega t}\; 0\; 0 \right]^T e^{-\gamma z}\,, \quad \mathbf{H_y}\, e^{-\gamma z} \Rightarrow \left[0\; (1/\eta) E_{x0} e^{j\omega t}\; 0 \right]^T e^{-\gamma z}$$

propagation constant: $\gamma = \alpha + j\beta = \alpha + j\omega / v_p = \alpha + j2\pi / \lambda = j\omega\sqrt{\mu\varepsilon}$

wavelength : $\lambda = 2\pi / \beta = 2\pi v_p / \omega = v_p / f$,

wavenumber (or spatial frequency): $k = \beta = \omega / v_p = 2\pi / \lambda$

intrinsic impedance : $\eta = j\omega\mu / \gamma = \sqrt{\mu / \varepsilon}$

phase velocity: $v_p = \dfrac{\omega}{\beta} = \left(\mu\varepsilon\right)^{-1/2}$

group velocity: $v_g = \dfrac{d\omega}{d\beta} = \left(\dfrac{d\beta}{d\omega}\right)^{-1} = \left(\mu\varepsilon\right)^{-1/2}$

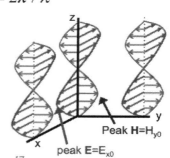

Peak H=H_{y0}

peak E=E_{x0}

Summary: Plane Wave Solution in Vacuum

wave equation: $\nabla^2 \mathbf{E} = \mu_0 \varepsilon_0 \dfrac{\partial^2 \mathbf{E}}{\partial t^2}$;

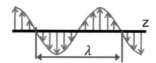

plane wave solution:

$$\mathbf{E_x}\, e^{-\gamma z} \Rightarrow \left[E_{x0} e^{j\omega t}\, 0\, 0 \right]^T e^{-\gamma z}, \quad \mathbf{H_y}\, e^{-\gamma z} \Rightarrow \left[0 \; (1/\eta) E_{x0} e^{j\omega t}\; 0 \right]^T e^{-\gamma z}$$

prop. const.: $\gamma = \alpha + j\beta = \alpha + j\omega/v_p = \alpha + j2\pi/\lambda = j\omega\sqrt{\mu_0\varepsilon_0} = j\omega/c$

wavelength: $\lambda = 2\pi/\beta = c/f$,

wavenumber: $k = \beta = \omega/c = 2\pi/\lambda$

intrinsic impedance : $\eta = j\omega\mu_0/\gamma = \sqrt{\mu_0/\varepsilon_0} = 377\ \Omega$

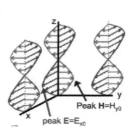

phase velocity: $v_p = \dfrac{\omega}{\beta} = (\mu\varepsilon)^{-1/2} = c = 3\times10^8$

group velocity: $v_g = \dfrac{d\omega}{d\beta} = \left(\dfrac{d\beta}{d\omega}\right)^{-1} = (\mu\varepsilon)^{-1/2} = c$

Peak H=H$_{y0}$

peak E=E$_{x0}$

Example: Plane Wave Solution

- Find the plane-wave parameters at 1 GHz for a printed-circuit board with ε_r=4.4 and μ_r=1

plane wave solution:

$$\mathbf{E_x}\, e^{-\gamma z} \Rightarrow \left[E_{x0} e^{j\omega t}\, 0\, 0 \right]^T e^{-\gamma z}, \quad \mathbf{H_y}\, e^{-\gamma z} \Rightarrow \left[0 \; (1/\eta) E_{x0} e^{j\omega t}\; 0 \right]^T e^{-\gamma z}$$

prop. const. $\gamma = j\omega\sqrt{\mu_r\mu_0\varepsilon_r\varepsilon_0}$

$$= j2\pi\times10^9\sqrt{(1)(1257\times10^{-9})(4.4)(8.85\times10^{-12})} = j44\ \text{rad/m}$$

wavelength: $\lambda = 2\pi/\beta = 2\pi/44 = 0.14$ m

wavenumber: $k = \beta = 2\pi/\lambda = 44$ rad/m

intrinsic impedance : $\eta = \sqrt{\mu/\varepsilon} = \sqrt{(1)(1257\times10^{-9})/[(4.4)(8.85\times10^{-12})]} = 179.6\ \Omega$

phase vel.: $v_p = (\mu\varepsilon)^{-1/2} = (\mu_r\mu_0\varepsilon_r\varepsilon_0)^{-1/2}$ Note this

$$= \dfrac{1}{\sqrt{(1)(1257\times10^{-9})(4.4)(8.85\times10^{-12})}} = 1.43\times10^8\ \text{m/s} = \dfrac{c}{\sqrt{\mu_r\varepsilon_r}}$$

group velocity: $v_g = (\mu\varepsilon)^{-1/2} = 1.43\times10^8$ m/s

Summary: Plane Wave Solution (Lossy Dielectric)

wave equation: $\nabla^2 \mathbf{E} = \mu\varepsilon \dfrac{\partial^2 \mathbf{E}}{\partial t^2}$;

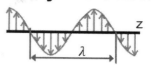

plane wave solution:

$$\mathbf{E}_x \, e^{-\gamma z} \Rightarrow \left[E_{x0} e^{j\omega t} \, 0 \, 0 \right]^T e^{-\gamma z} \ , \quad \mathbf{H}_y \, e^{-\gamma z} \Rightarrow \left[0 \ (1/\eta) E_{x0} e^{j\omega t} \ 0 \right]^T e^{-\gamma z}$$

propagation constant: $\gamma = \alpha + j\beta = j\omega\sqrt{\mu\varepsilon}\sqrt{1 - j\sigma/(\omega\varepsilon)}$

wavelength : $\lambda = 2\pi/\beta = 2\pi/\mathrm{Im}(\gamma)$,

Note: $\gamma = \alpha + j\beta$ where α is no longer zero

wavenumber (or spatial frequency): $k = \beta = \mathrm{Im}(\gamma)$

intrinsic impedance : $\eta = j\omega\mu/\gamma = \sqrt{\mu/\varepsilon}\left(1 - j\sigma/(\omega\varepsilon)\right)^{-1/2}$

phase velocity: $v_p = \dfrac{\omega}{\beta}$

group velocity: $v_g = \dfrac{d\omega}{d\beta} = \left(\dfrac{d\beta}{d\omega}\right)^{-1}$

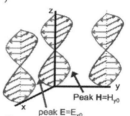

Peak H=H$_{y0}$

peak E=E$_{x0}$

Example: Lossy Plane Wave Solution

- Find plane-wave parameters at 1 GHz for a material with ε_r=4.4, μ_r=1, and σ= 0.0024 S/m, and find loss in dB at 3 m

$$\mathbf{E}_x \, e^{-\gamma z} \Rightarrow \left[E_{x0} e^{j\omega t} \, 0 \, 0 \right]^T e^{-\gamma z} \ , \quad \mathbf{H}_y \, e^{-\gamma z} \Rightarrow \left[0 \ (1/\eta) E_{x0} e^{j\omega t} \ 0 \right]^T e^{-\gamma z}$$

prop. const. $\gamma = \alpha + j\beta = j\omega\sqrt{\mu_r\mu_0\varepsilon_r\varepsilon_0}\sqrt{1 - j\sigma/(\omega\varepsilon_r\varepsilon_0)}$

$= j2\pi \times 10^9 \sqrt{(1)(1257 \times 10^{-9})(4.4)(8.85 \times 10^{-12})}\sqrt{1 - \dfrac{j0.0024}{(2\pi \times 10^9)(1)(8.85 \times 10^{-12})}} = 0.22 + j44$ rad/m

wavelength: $\lambda = 2\pi/\beta = 2\pi/44 = 0.14$ m

wavenumber: $k = \beta = 2\pi/\lambda = 44$ rad/m

attenuation constant: $\alpha = 0.22$ Neper/m,

loss at 3 m: $10\log_{10}\left((e^{-\alpha z})^2\right) = 10\log_{10}\left((e^{-(0.22)(3)})^2\right) = -5.7$ dB $\quad \Rightarrow$ loss=5.7 dB

intrinsic impedance : $\eta = j\omega\mu/\gamma = \dfrac{2\pi \times 10^9}{0.22 + j44} = 180 + j0.9\ \Omega$

phase vel.: $v_p = \omega/\beta = (2\pi \times 10^9)/44 = 1.43 \times 10^8$ m/s

group velocity: $v_g = \dfrac{d\omega}{d\beta} = \left(\dfrac{d\beta}{d\omega}\right)^{-1} = 1.43 \times 10^8$ m/s (computed)

Alternative e^{-jkz} Form

Not Used by Us,
But In Many Textbooks

Plane Wave Solution, e^{-jkz} Form

$$\nabla^2 \mathbf{E} = -\mu\varepsilon \frac{\partial^2 \mathbf{E}}{\partial t^2}$$

$$\nabla^2 \mathbf{E} = \left(\frac{\partial^2}{\partial x^2} + \frac{\partial^2}{\partial y^2} + \frac{\partial^2}{\partial z^2} \right) \mathbf{E} = -\mu\varepsilon \frac{\partial^2 \mathbf{E}}{\partial t^2}$$

- The preceding plane-wave solution is similar to our transmission line form, but many textbooks use an e^{-jkz} form
- So here we repeat the solution using an e^{-jkz} form
  - The E-field vector only contains x-axis components
  - The wave only travels in z-axis direction (phase velocity v_p)
  - The region is source-free as before, $\nabla \cdot \mathbf{D} = \rho_e = 0$, $\mathbf{J}_e = 0$
- As for the "A e$^{-\gamma z}$ e$^{j\omega t}$" form, guess that the solution has form of "A e^{-jkz} e$^{j\omega t}$"
- Guess:

$$\text{guess: } \mathbf{E} = E_{x0} e^{-jkz} e^{j\omega t} \begin{bmatrix} 1 \\ 0 \\ 0 \end{bmatrix} = \begin{bmatrix} E_{x0} e^{-jkz} e^{j\omega t} \\ 0 \\ 0 \end{bmatrix}$$

Wave only travels in z-direction
since Re{e$^{-\gamma z}$ e$^{j\omega t}$}=cos(ωt-γz)

Only has x-component

Plane Wave Solution, e^{-jkz} Form

$$\nabla^2 \mathbf{E} = -\mu\varepsilon \frac{\partial^2 \mathbf{E}}{\partial t^2}$$

$$\nabla^2 \mathbf{E} = \left(\frac{\partial^2}{\partial x^2} + \frac{\partial^2}{\partial y^2} + \frac{\partial^2}{\partial z^2} \right) \mathbf{E} = -\mu\varepsilon \frac{\partial^2 \mathbf{E}}{\partial t^2}$$

- Using plane-wave solution guess from previous slide:

$$\left(\frac{\partial^2}{\partial x^2} + \frac{\partial^2}{\partial y^2} + \frac{\partial^2}{\partial z^2} \right) \begin{bmatrix} E_{x0} e^{-jkz} e^{j\omega t} \\ 0 \\ 0 \end{bmatrix} = \mu\varepsilon \frac{\partial^2 \mathbf{E}}{\partial t^2} = \mu\varepsilon \frac{\partial^2}{\partial t^2} \begin{bmatrix} E_{x0} e^{-jkz} e^{j\omega t} \\ 0 \\ 0 \end{bmatrix}$$

$$\Rightarrow \frac{\partial^2}{\partial z^2} \begin{bmatrix} E_{x0} e^{-jkz} e^{j\omega t} \\ 0 \\ 0 \end{bmatrix} = \mu\varepsilon \frac{\partial^2}{\partial t^2} \begin{bmatrix} E_{x0} e^{-jkz} e^{j\omega t} \\ 0 \\ 0 \end{bmatrix}$$

$$\Rightarrow \frac{\partial^2 E_{x0} e^{-jkz} e^{j\omega t}}{\partial z^2} = \mu\varepsilon \frac{\partial^2 E_{x0} e^{-\gamma z} e^{j\omega t}}{\partial t^2}$$

$$\Rightarrow -k^2 E_{x0} e^{-\gamma z} e^{j\omega t} = -\omega^2 \mu\varepsilon E_{x0} e^{-\gamma z} e^{j\omega t}$$

$$so: k^2 = \omega^2 \mu\varepsilon, \quad or \quad k = \omega\sqrt{\mu\varepsilon}, \quad \pm k \text{ are solutions}$$

- Which gives solution for wavenumber k

Final Plane Wave Solution, e^{-jkz} Form

$$\nabla^2 \mathbf{E} = -\mu\varepsilon \frac{\partial^2 \mathbf{E}}{\partial t^2}$$

$$\nabla^2 \mathbf{E} = \left(\frac{\partial^2}{\partial x^2} + \frac{\partial^2}{\partial y^2} + \frac{\partial^2}{\partial z^2} \right) \mathbf{E} = -\mu\varepsilon \frac{\partial^2 \mathbf{E}}{\partial t^2}$$

- Finally, the plane-wave solution from previous slide is:

Exact same solution as with $e^{-\gamma z}$ form

$$with: k^2 = \omega^2 \mu\varepsilon, \quad or \quad k = \pm j\omega\sqrt{\mu\varepsilon}$$

$$\mathbf{E} = \begin{bmatrix} E_{x0} e^{-jkz} e^{j\omega t} \\ 0 \\ 0 \end{bmatrix} = E_{x0} e^{-jkz} e^{j\omega t} \begin{bmatrix} 1 \\ 0 \\ 0 \end{bmatrix} = E_{x0} e^{\pm j\omega\sqrt{\mu\varepsilon}\, z} e^{j\omega t} \begin{bmatrix} 1 \\ 0 \\ 0 \end{bmatrix}$$

- And we will adopt a phasor shorthand "$E_x e^{-jkz}$" defined as

$$definition: \mathbf{E}_x \, e^{-jkz} \Rightarrow \begin{bmatrix} E_{x0} e^{j\omega t} \\ 0 \\ 0 \end{bmatrix} e^{-jkz}, \; or \; \mathbf{E}\, e^{-jkz} = \begin{bmatrix} E_{x0} e^{j\omega t} \\ E_{y0} e^{j\omega t} \\ E_{z0} e^{j\omega t} \end{bmatrix} e^{-jkz}$$

- Beware: different authors use different phasor notations, such as Hayt's $\mathbf{E}_{sx}$, and Ellingson $\tilde{\mathbf{E}}_x$, and Ida $E_x(z)$

Comparison: Plane Wave Forms e^{-jkz} and e$^{-\gamma z}$

$$\nabla^2 \mathbf{E} = -\mu\varepsilon \frac{\partial^2 \mathbf{E}}{\partial t^2}$$

- There are two commonly-used forms of plane-wave solutions: "A e$^{-\gamma z}$ e$^{j\omega t}$" and "A e^{-jkz} e$^{j\omega t}$"

- We can see the relationship by setting A e$^{-\gamma z}$ e$^{j\omega t}$ = A e^{-jkz} e$^{j\omega t}$, then it is clear that γ = jk , so:

  Beware the minus sign

  o Im(γ) = β = Re(k)

  o Re(γ) = α = -Im(k)

- Both forms resulted in the same solution

- Also note that:

  o Re(A e$^{-\gamma z}$ e$^{j\omega t}$)=A e$^{-Re(\gamma)z}$ cos(ωt - Im(γ)z)

  o Re(A e^{-jkz} e$^{j\omega t}$)=A e$^{+Im(k)z}$ cos(ωt − Re(k)z)

- From this point forward, we will primarily use an e^{-jkz} form

7 PLANE WAVES AT BOUNDARIES

The lecture notes in this chapter discuss the behavior of electromagnetic waves at boundariess.

Boundary Conditions

Electrostatic Boundary Conditions

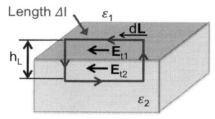
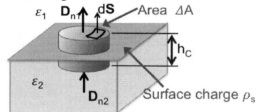

- Above shows the interface between two dielectrics ε_1 and ε_2
- By Kirchoff's law around the loop on the left, as $h_L \rightarrow 0$:

$$0 = \lim_{h_L \to 0} \oint_L \mathbf{E} \cdot d\mathbf{L} = \Delta l \left(\mathbf{E}_{t1} - \mathbf{E}_{t2} \right) \Rightarrow \boxed{\text{so} \quad \mathbf{E}_{t1} = \mathbf{E}_{t2}}$$

- By Gauss' law around the surface of the cylinder, as $h_C \rightarrow 0$:

$$Q = \lim_{h_C \to 0} \int_V \rho_e \, dv = \rho_S \Delta A = \lim_{h_C \to 0} \oint_S \mathbf{D} \cdot d\mathbf{S} = \Delta A \left(\mathbf{D}_{n1} - \mathbf{D}_{n2} \right)$$

$$\Rightarrow \boxed{\text{so} \quad \mathbf{D}_{n1} - \mathbf{D}_{n2} = \varepsilon_1 \mathbf{E}_{n1} - \varepsilon_2 \mathbf{E}_{n2} = \rho_S}$$

Magnetostatic Boundary Conditions

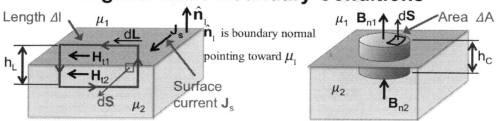

- Above shows interface of two magnetic materials μ_1 and μ_2
- By Ampere circuital law around the loop on the left, as $h_L \to 0$:

$$I = \lim_{h_L \to 0} \int_S \mathbf{J_e} \cdot d\mathbf{S} = \Delta l \mathbf{J_s} \cdot \hat{\mathbf{n}}_S = \lim_{h_L \to 0} \oint_L \mathbf{H} \cdot d\mathbf{L} = \Delta l \left(\mathbf{H}_{t1} - \mathbf{H}_{t2} \right) \cdot \hat{\mathbf{i}}$$

$$\Rightarrow \quad \boxed{\text{so} \quad \mathbf{H}_{t1} - \mathbf{H}_{t2} = \mathbf{J_s} \times \hat{\mathbf{n}}_1} \quad \text{where } \hat{\mathbf{n}}_1 \text{ is normal pointing toward } \mu_1$$

- By Gauss' law around the surface of the cylinder, as $h_C \to 0$:

$$0 = \lim_{h_C \to 0} \oint_S \mathbf{B} \cdot d\mathbf{S} = \Delta A \left(\mathbf{B}_{n1} - \mathbf{B}_{n2} \right) \quad \Rightarrow \quad \boxed{\text{so} \quad \mathbf{B}_{n1} = \mathbf{B}_{n2} \text{ or } \mu_1 \mathbf{H}_{n1} = \mu_2 \mathbf{H}_{n2}}$$

Time-Varying Boundary Conditions

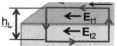 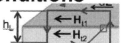

- For our purposes and most typical situations, the time-varying boundary conditions are the same as the foregoing electrostatic and magnetostatic boundary conditions
- This is true, since only the two tangential (Kirchoff and Ampere) cases would be affected by d/dt terms
- For Kirchoff (Faraday) as $h_L \to 0$, presuming no "sheet B"

$$\lim_{h_L \to 0} \int_S \frac{-\partial \mathbf{B}}{\partial t} \cdot d\mathbf{S} = 0 = \lim_{h_L \to 0} \oint_L \mathbf{E} \cdot d\mathbf{L} = \Delta l \left(\mathbf{E}_{t1} - \mathbf{E}_{t2} \right) \quad \Rightarrow \quad \boxed{\text{so} \quad \mathbf{E}_{t1} = \mathbf{E}_{t2}}$$

- Similarly, as $h_L \to 0$ for Ampere, presuming no "sheet D"

$$I = \lim_{h_L \to 0} \int_S \left(\mathbf{J_e} + \frac{\partial \mathbf{D}}{\partial t} \right) \cdot d\mathbf{S} = \Delta l \mathbf{J_s} \cdot \hat{\mathbf{n}}_S = \lim_{h_L \to 0} \oint_L \mathbf{H} \cdot d\mathbf{L} = \Delta l \left(\mathbf{H}_{t1} - \mathbf{H}_{t2} \right)$$

$$\Rightarrow \quad \boxed{\text{so} \quad \mathbf{H}_{t1} - \mathbf{H}_{t2} = \mathbf{J_s} \times \hat{\mathbf{n}}_1} \quad \text{where } \hat{\mathbf{n}}_1 \text{ is normal pointing toward } \mu_1$$

Time-Varying Boundary Conditions Dependency

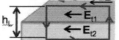 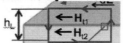

- For time-varying boundary conditions, it can be shown that the tangential field conditions are sufficient, because the two Maxwell divergence relations can be derived from the Maxwell curl relations
  from Maxwell: $\quad \nabla \times \mathbf{E} = -\dfrac{\partial \mathbf{B}}{\partial t} \quad$ and $\quad \nabla \times \mathbf{H} = \mathbf{J}_e + \dfrac{\partial \mathbf{D}}{\partial t}$

  taking divergence, and noting vector identity $\nabla \cdot (\nabla \times \mathbf{E}) = 0$

  $$\nabla \cdot (\nabla \times \mathbf{E}) = 0 = -\frac{\partial \nabla \cdot \mathbf{B}}{\partial t} \quad \text{and} \quad \nabla \cdot (\nabla \times \mathbf{H}) = 0 = \nabla \cdot \mathbf{J}_e + \frac{\partial \nabla \cdot \mathbf{D}}{\partial t}$$

  integrating over time:

  $$\boxed{0 = \nabla \cdot \mathbf{B}} \quad \text{and} \quad 0 = \int \nabla \cdot \mathbf{J}_e \, dt + \nabla \cdot \mathbf{D} \Rightarrow \boxed{0 = \rho_e + \nabla \cdot \mathbf{D}}$$

- Thus, 2 of Maxwell's equations can be derived from Faraday and Ampere, for the time-varying case

Credit: Ramo, Whinnery, VanDuzer, "Fields and Waves in Comm. Electronics," JWiley&Sons,1984.

Perfect Electric Conductor (PEC) Boundary

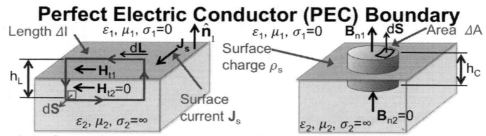

- Interface of conductor, $\sigma_2 = \infty$, with non-conductor ε_1, μ_1, $\sigma_1 = 0$
- Inside perfect conductor: $\mathbf{E} = 0$, $\mathbf{H} = 0$
- Surface may have surface current $\mathbf{J}_S$ and/or surface charge ρ_s
- For electric boundary condition with $\mathbf{E} = 0$ inside conductor:
  $$\mathbf{E}_{t1} = \mathbf{E}_{t2} \Rightarrow \text{ so } \boxed{\mathbf{E}_{t1} = 0} \text{ and } \mathbf{D}_{n1} - \mathbf{D}_{n2} = \rho_S \Rightarrow \text{ so } \boxed{\mathbf{D}_{n1} = \rho_S}$$
- For magnetic boundary condition with $\mathbf{H} = 0$ inside conductor:
  $$\mathbf{B}_{n1} = \mathbf{B}_{n2} \Rightarrow \text{ so } \boxed{\mathbf{B}_{n1} = 0}$$
  $$\mathbf{H}_{t1} - \mathbf{H}_{t2} = \mathbf{J_s} \times \hat{\mathbf{n}}_1 \Rightarrow \text{so} \boxed{\mathbf{H}_{t1} = \mathbf{J_s} \times \hat{\mathbf{n}}_1} \text{ where } \hat{\mathbf{n}}_1 \text{ is normal pointing toward } \mu_1$$
- Above applies to both static and time-varying cases

Plane Wave at Normal Incidence to Boundaries

Plane Wave at Normal Incidence to Dielectric

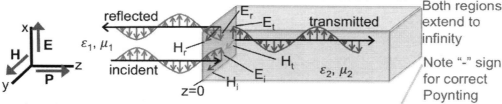

Both regions extend to infinity

Note "-" sign for correct Poynting

- As shown, a vertically polarized incident plane wave travels along z-axis in material1 with ε_1, μ_1, in phasor form $\mathbf{E_{xi}} = E_{i0}\, e^{-\gamma_1 z}$ and $\mathbf{H_{yi}} = H_{yi0} e^{-\gamma_1 z}$ hits a boundary at z=0 of material2 with ε_2, μ_2
- Reflected is $\mathbf{E_{xr}} = [\, E_{xr0}\ 0\ 0\,]^T e^{+\gamma_1 z}$, and $\mathbf{H_{yr}} = [\,0\ -E_{yr0}/\eta_1\ 0\,]^T e^{+\gamma_1 z}$
- Transmitted is $\mathbf{E_{xt}} = [\, E_{xt0}\ 0\ 0\,]^T e^{-\gamma_2 z}$, and $\mathbf{H_{xt}} = [\,0\ E_{xt0}/\eta_2\ 0\,]^T e^{-\gamma_2 z}$
- At the boundary z=0, $E_{t1} = E_{t2}$, and $H_{t1} = H_{t2}$ for $J_S = 0$, so:

$$\mathbf{E_{xi0}}\, e^{-\gamma_1(0)} + \mathbf{E_{xr0}}\, e^{+\gamma_1(0)} = \mathbf{E_{xt0}}\, e^{-\gamma_2(0)} \Rightarrow E_{xi0} + E_{xr0} = E_{xt0}$$

$$\mathbf{H_{yi0}} + \mathbf{H_{yr0}} = \mathbf{H_{yt0}} \Rightarrow H_{yi0} + H_{xr0} = (E_{xi0} - E_{xr0})/\eta_1 = H_{xt0} = E_{xt0}/\eta_2$$

solving: $\boxed{\dfrac{E_{xr0}}{E_{xi0}} = \dfrac{\eta_2 - \eta_1}{\eta_1 + \eta_2} = \Gamma,}$ and $\boxed{\dfrac{E_{xt0}}{E_{xi0}} = 1 + \Gamma = \dfrac{2\eta_2}{\eta_1 + \eta_2}}$ ←Note: same as RC line

Example: Plane Wave Normal Incidence to Dielectric

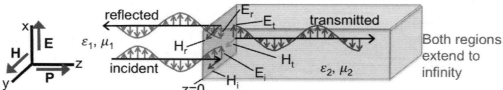

- Example: vertically polarized incident plane wave in material 1 having $\mathbf{E}_{xi}=[\,E_{xi0}\,0\,0\,]^{T}e^{-\gamma_1 z}$, and $\mathbf{H}_{yi}=[\,0\,-E_{yi0}/\eta_1\,0\,]^{T}e^{-\gamma_1 z}$ travels along z-axis with $\varepsilon_1=1$, $\mu_1=1$, and impinges on a boundary at z=0 with material 2 with $\varepsilon_2=4$, $\mu_2=1$, find reflected and incident

$$\gamma_1 = j\omega\sqrt{\mu\varepsilon} = j\omega\sqrt{\mu_0\varepsilon_0} = j\omega/c, \quad \eta_1 = \sqrt{\mu/\varepsilon} = \sqrt{\mu_0/\varepsilon_0} = 377$$

$$\gamma_2 = j\omega\sqrt{\mu_0 4\varepsilon_0} = j2\omega/c, \quad \eta_2 = \sqrt{\mu/\varepsilon} = \sqrt{\mu_0/(4\varepsilon_0)} \approx 188$$

Note "-" sign for correct Poynting

$$\Gamma = (\eta_2 - \eta_1)/(\eta_1 + \eta_2) = (189-377)/(377+189) = -1/3 = e^{-j\pi}/3$$

$$\mathbf{E}_{xr} = \Gamma\mathbf{E}_{xi0}e^{+\gamma_1 z} = (e^{-j\pi}/3)\big[E_{xi0}\,0\,0\big]^{T}e^{+j\omega z/c} = \big[(E_{xi0}/3)\,0\,0\big]^{T}e^{j(\omega z/c - \pi)}$$

$$\mathbf{E}_{xt} = (1+\Gamma)\mathbf{E}_{xi0}e^{-\gamma_2 z} = (2/3)\big[E_{xi0}\,0\,0\big]^{T}e^{-j2\omega z/c} = \big[(2E_{xi0}/3)\,0\,0\big]^{T}e^{-j2\omega z/c}$$

$$H_{yr0} = \frac{-E_{xr0}}{\eta_1} \Rightarrow \left[0\ \frac{-E_{xi0}}{1131}\ 0\right]^{T}e^{j(\omega z/c-\pi)} \quad H_{yto} = \frac{E_{xt0}}{\eta_2} \Rightarrow \left[0\ \frac{2E_{xi0}}{564}\ 0\right]^{T}e^{-j2\omega z/c}$$

Example: Power: Plane Wave Normal Incidence to Dielectric

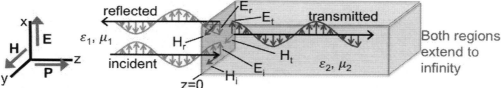

- Example: For the previous example, let the incident field be 100 V/m, find the E and H fields and average real power densities

$$\mathbf{E}_{xi} = \mathbf{E}_{xi0}e^{-\gamma_1 z} = \big[100\,0\,0\big]^{T}e^{-j\omega z/c}, \mathbf{H}_{xi} = \big[0\,(100/377)\,0\big]^{T}e^{j\omega z/c}$$

$$0.5 * \mathrm{Re}\big\{\mathbf{E}_{xi} \times \mathbf{H}_{xi}^{*}\big\} = 13.3\,\hat{\mathbf{z}}\ \text{W/m}^2$$

Note that power is conserved

$$\mathbf{E}_{xr} = \big[(100/3)\,0\,0\big]^{T}e^{j(\omega z/c-\pi)}, \mathbf{H}_{xr} = \left[0\ \frac{-100}{1131}\ 0\right]^{T}e^{j(\omega z/c-\pi)}$$

$$0.5 * \mathrm{Re}\big\{\mathbf{E}_{xr} \times \mathbf{H}_{xr}^{*}\big\} = -1.5\,\hat{\mathbf{z}}\ \text{W/m}^2$$

$$\mathbf{E}_{xt} = \big[(200E_{xi0}/3)\,0\,0\big]^{T}e^{-j2\omega z/c}, \mathbf{H}_{xt} = \left[0\ \frac{200}{564}\ 0\right]^{T}e^{-j2\omega z/c}$$

Note correct Poynting shows flow backward

$$0.5 * \mathrm{Re}\big\{\mathbf{E}_{xt} \times \mathbf{H}_{xt}^{*}\big\} = 11.8\,\hat{\mathbf{z}}\ \text{W/m}^2$$

Comparison To Transmission Line Model

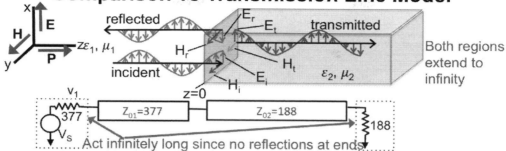

- The infinitely long pair of transmission lines in the circuit above models the plane wave

- For the transmission line circuit, at the interface:

$$\Gamma = \frac{V_{ref}}{V_{inc}} = \frac{Z_{02} - Z_{01}}{Z_{01} + Z_{02}}, \quad \text{and} \quad V_{trans} = (1+\Gamma)V_{inc} = \frac{2Z_{02}}{Z_{01} + Z_{02}}$$

- Which is the same as the plane-wave result, except replacing Z_{01} and Z_{02} by η_1 and η_2 by and replacing voltages by E-fields

$$\frac{E_{xr0}}{E_{xi0}} = \frac{\eta_2 - \eta_1}{\eta_1 + \eta_2} = \Gamma, \quad \text{and} \quad \frac{E_{xt0}}{E_{xi0}} = 1+\Gamma = \frac{2\eta_2}{\eta_1 + \eta_2}$$

Finite Thickness Transmission Line Model

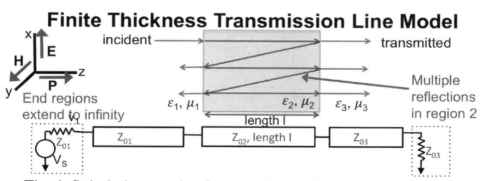

- The infinitely long pair of transmission lines with a third line of finite length between models the plane wave situation above

- The situation is more complicated due to multiple reflections

- Nevertheless, one simple case is when length l is a half wave, then the impedance presented to the incident wave is clearly Z_{03} and $\Gamma=(Z_{03}- Z_{01})/(Z_{03}+ Z_{01})$ or for the plane wave"

$$\text{For } l = \lambda/2, \quad E_{xr0} / E_{xi0} = \Gamma = (\eta_3 - \eta_1)/(\eta_1 + \eta_3)$$

Plane Wave at Normal Incidence to Conductor

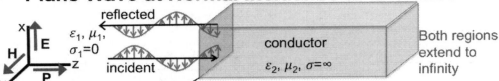

- As shown, a vertically polarized incident plane wave travels along z-axis in material1 with ε_1, μ_1, $\sigma_1=0$ in phasor form $\mathbf{E}_{xi}=$ $\mathbf{E}_{i0}\,e^{-\gamma 1z}$ and $\mathbf{H}_{yi}=\mathbf{H}_{yi0}e^{-\gamma 1z}$ with a conductor boundary at z=0
- No electric or magnetic field is inside the conductor
- At the boundary z=0, $E_{t1}=0$, and $H_{t1}=J_S$, so:

Note: same as for transmission line with shorted line

$$\gamma_1 = j\omega\sqrt{\mu_1\varepsilon_1}, \qquad \eta_1 = \sqrt{\mu_1/\varepsilon_1}$$

$$\mathbf{E}_{xi0}\,e^{-\gamma_1(0)} + \mathbf{E}_{xr0}\,e^{+\gamma_1(0)} = 0 \quad \Rightarrow E_{xr0} = -E_{xi0}, \Rightarrow \boxed{\Gamma = E_{xr0}/E_{xi0} = -1}$$

$$\mathbf{E}_{xi} = \left[E_{xi0}\ 0\ 0\right]^T e^{-\gamma_1 z}, \mathbf{H}_{yi} = \left[0\ (E_{xi0}/\eta_1)\ 0\right]^T e^{-\gamma_1 z}$$

$$\mathbf{E}_{xr} = \left[-E_{xi0}\ 0\ 0\right]^T e^{+\gamma_1 z}, \mathbf{H}_{yr} = \left[0\ (E_{xi0}/\eta_1)\ 0\right]^T e^{+\gamma_1 z}$$

Note: $H \neq 0$ @z=0

Note: at interface at z=0: $\mathbf{H}_{ytot} = \mathbf{H}_{yi} + \mathbf{H}_{yr} = \left[0\ (2E_{xi0}/\eta_1)\ 0\right]^T \neq 0$

Plane Wave at Oblique Incidence to Boundaries

Lossless Plane Waves On X- Axis and Wavevector K

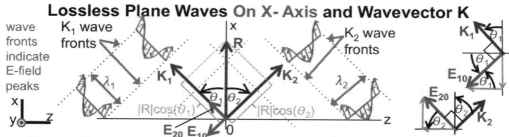

- Above, 2 plane waves: one in direction of wavevector K_1 to the left, one in direction K_2, and H-fields point out of page for both, E-fields orthogonal to K_1 and K_2, and $|K_1|=\beta_1$ and $|K_2|=\beta_2$ ←If lossless

- The lossless plane wave E-fields at location R on the x-axis are

$$\mathbf{E}_{K1} = \mathbf{E}_{10}\, e^{-j\mathbf{K1\cdot R}} \Rightarrow |\mathbf{E}_{10}|\left[-\sin\left(\theta_1\right)\ 0\ -\cos\left(\theta_1\right)\right]^T e^{j\omega t} e^{-j\beta_1 d_1},\ \text{for } d_1 = |\mathbf{R}|\cos\left(\theta_1\right),\ \beta_1 = |\mathbf{K}_1|$$

$$\mathbf{E}_{K2} = \mathbf{E}_{20}\, e^{-j\mathbf{K2\cdot R}} \Rightarrow |\mathbf{E}_{20}|\left[\sin\left(\theta_2\right)\ 0\ -\cos\left(\theta_2\right)\right]^T e^{j\omega t} e^{-j\beta_2 d_2},\ \text{for } d_2 = |\mathbf{R}|\cos\left(\theta_2\right),\ \beta_2 = |\mathbf{K}_2|$$

- For the example shown with position vector **R** on the x-axis, at **R** the phase of the plane waves are given by the dot products
  $\mathbf{R\cdot K_1}=|\mathbf{R}||\mathbf{K_1}|\cos(\theta_1)=|\mathbf{R}||\beta_1|\cos(\theta_1)=|\mathbf{R}|(2\pi/\lambda_1)\cos(\theta_1)$ and
  $\mathbf{R\cdot K_2}=|\mathbf{R}||\mathbf{K_2}|\cos(\theta_2)=|\mathbf{R}||\beta_2|\cos(\theta_2)=|\mathbf{R}|(2\pi/\lambda_2)\cos(\theta_2)$

Lossless Plane Waves Off Axis at Arbitrary Location R

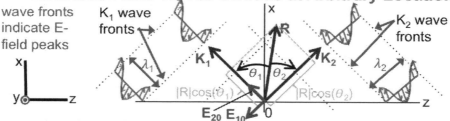

- As in prior slide, 2 plane waves in direction of wavevectors K_1 $=\beta_1[\cos(\theta_1)\ 0\ -\sin(\theta_1)]$ to left and K_2 to right, H-field out of page

- For arbitrary position vector **R**, at location **R** the phase of the two plane waves are given by the dot products

$$\mathbf{E}_{K1} \Rightarrow |\mathbf{E}_{10}|\left[-\sin(\theta_1)\ 0\ -\cos(\theta_1)\right]^T e^{j\omega t} e^{-j\mathbf{K_1\cdot R}} = |\mathbf{E}_{10}|\left[-\sin(\theta_1)\ 0\ -\cos(\theta_1)\right]^T e^{j\omega t} e^{-j|\mathbf{R}||\mathbf{K}_1|\cos(\theta_1)}$$

$$\mathbf{E}_{K2} \Rightarrow |\mathbf{E}_{20}|\left[\sin(\theta_2)\ 0\ -\cos(\theta_2)\right]^T e^{j\omega t} e^{-j\mathbf{K_2\cdot R}} = |\mathbf{E}_{20}|\left[\sin(\theta_2)\ 0\ -\cos(\theta_2)\right]^T e^{j\omega t} e^{-j|\mathbf{R}||\mathbf{K}_2|\cos(\theta_2)}$$

- Above, the phase of the plane wave of K_1 at location **R** is near a wave front peak of the E-field, and phase is determined by
  $\mathbf{E}_{10}e^{-j\mathbf{R\cdot K1}} =\mathbf{E}_{10}e^{-j|\mathbf{R}||\mathbf{K1}|\cos(\theta1)} =\mathbf{E}_{10}e^{-j|\mathbf{R}|\beta1\cos(\theta1)} =\mathbf{E}_{10}e^{-j|\mathbf{R}|(2\pi/\lambda1)\cos(\theta1)}$

- And for K_2, $\mathbf{E}_{20}e^{-j\mathbf{R\cdot K2}} =e^{-j|\mathbf{R}||\beta2|\cos(\theta2)}=e^{-j|\mathbf{R}|(2\pi/\lambda2)\cos(\theta2)}$

Time=Δt: Lossless Plane Waves and Wavevector K

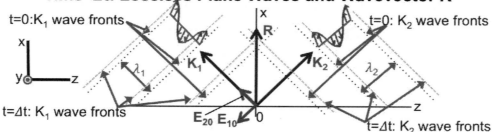

- Wave fronts move slightly as time increases slightly by Δt
- The wave fronts are "E-field peaks," corresponding to points where Re$\{|E_0| e^{j(\omega t+\theta)} e^{-j\beta z}\}$=$|E_0|\cos(\omega t+\theta-\beta z)$=$|E_0|\cos(n2\pi)$=$|E_0|$
- Prior slides were "snapshots" of wave fronts at time t=0
- At some small time instant Δt later, at t=Δt, the two plane waves will have moved slightly, such that their wave fronts also move slightly in then direction of their wavevector K_1 and K_2,
- They may move different distances if their wavevector magnitudes are not equal, since $|K_1|=\beta_1=\omega/v_{p1}$ and $|K_2|=\omega/v_{p2}$

Two Plane Waves: Cancel E-field X-Component on X-Axis

Note: as drawn, x-components of E_{10} and E_{20} would cancel

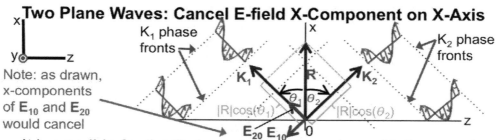

- It is possible for the 2 plane waves above to perfectly cancel the x-component of the E-field at every point on the x-axis
- The two plane wave E-fields are

$$\mathbf{E}_{K1} \Rightarrow |\mathbf{E}_{10}|\left[-\sin(\theta_1)\ 0\ -\cos(\theta_1)\right]^T e^{j\omega t}e^{-j\mathbf{K}_1\cdot\mathbf{R}} = |\mathbf{E}_{10}|\left[-\sin(\theta_1)\ 0\ -\cos(\theta_1)\right]^T e^{j\omega t}e^{-j|\mathbf{R}||\mathbf{K}_1|\cos(\theta_1)}$$

$$\mathbf{E}_{K2} \Rightarrow |\mathbf{E}_{20}|\left[\sin(\theta_2)\ 0\ -\cos(\theta_2)\right]^T e^{j\omega t}e^{-j\mathbf{K}_2\cdot\mathbf{R}} = |\mathbf{E}_{20}|\left[\sin(\theta_2)\ 0\ -\cos(\theta_2)\right]^T e^{j\omega t}e^{-j|\mathbf{R}||\mathbf{K}_2|\cos(\theta_2)}$$

- To cancel the x-component at z=0 on the x-axis

$$0 = |\mathbf{E}_{20}|\sin(\theta_2)e^{j\omega t}e^{-j|\mathbf{R}||\mathbf{K}_2|\cos(\theta_2)} - |\mathbf{E}_{10}|\sin(\theta_1)^{j\omega t}e^{-j|\mathbf{R}||\mathbf{K}_1|\cos(\theta_1)}$$

Note the two H_y-fields <u>reinforce</u> as drawn so total $H_{y0}=H_{y10}+H_{y20}$

$$\Rightarrow |\mathbf{E}_{20}|\sin(\theta_2)e^{-j|\mathbf{R}|\beta_2\cos(\theta_2)} = |\mathbf{E}_{10}|\sin(\theta_1)e^{-j|\mathbf{R}|\beta_1\cos(\theta_1)}$$

$$\boxed{\Rightarrow |\mathbf{E}_{10}| = |\mathbf{E}_{20}|\ \text{and}\ \theta_1=\theta_2\ \text{and}\ \beta_1=\beta_2\ \text{for zero x-component of E at z=0}}$$

Plane Wave at Oblique Incidence to Conductors

TM Plane Wave at Oblique Incidence to Conductor

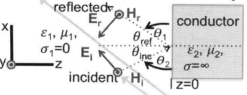
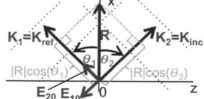

- As shown, a plane wave with wavevector $\mathbf{K}_2$ is incident on a conductor with boundary in the x-y plane and **H** out of the page
- The plane of incidence (the page) contains the normal to the boundary and the two plane-wave wavevectors
- No electric or magnetic field is inside the conductor
- Same as prior slide, with $\mathbf{K}_2$ incident, $\mathbf{K}_1$ reflected wavevectors
- At the boundary z=0, $E_{t1}=0$,

$$0 = \left|\mathbf{E}_{20}\right|\sin(\theta_2)e^{j\omega t}e^{-j|\mathbf{R}||\mathbf{K}_2|\cos(\theta_2)} - \left|\mathbf{E}_{10}\right|\sin(\theta_1)e^{j\omega t}e^{-j|\mathbf{R}||\mathbf{K}_1|\cos(\theta_1)}$$

$$\Rightarrow \left|\mathbf{E}_{20}\right|\sin(\theta_2)e^{-j|\mathbf{R}|\beta_2\cos(\theta_2)} = \left|\mathbf{E}_{10}\right|\sin(\theta_1)e^{-j|\mathbf{R}|\beta_1\cos(\theta_1)}$$

$$\boxed{\Rightarrow \left|\mathbf{E}_{10}\right| = \left|\mathbf{E}_{20}\right| \quad and \quad \beta_1 = \beta_2 \quad and \quad \theta_1 = \theta_2 \Rightarrow \theta_{inc} = \theta_{ref}}$$

"TM polarization" because magnetic field is transverse to incidence plane

Note: $H_y \ne 0$ @z=0

TM H-Fields at Oblique Incidence to Conductor

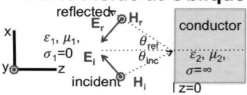
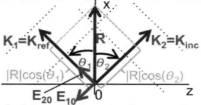

- As shown for the TM case, the H-fields point out of page for both the incident and reflected waves,

- The H-fields can the be computed from the E-fields as:
$$|\mathbf{E}_{20}| = |\mathbf{E}_{20}| = |\mathbf{E}| \quad and \quad \theta_1 = \theta_2$$

$$\mathbf{E}_{K1} \Rightarrow |\mathbf{E}_{10}| \left[-\sin(\theta_1)\ 0\ -\cos(\theta_1) \right]^T e^{j\omega t} e^{-j|\mathbf{R}||\mathbf{K}_1|\cos(\theta_1)}$$

$$\Rightarrow \mathbf{H}_{K1} = \frac{|\mathbf{E}|}{\eta_1} \left[0\ 1\ 0 \right]^T e^{j\omega t} e^{-j|\mathbf{R}||\mathbf{K}_1|\cos(\theta_1)}$$

$$\mathbf{E}_{K2} \Rightarrow |\mathbf{E}_{20}| \left[-\sin(\theta_2)\ 0\ -\cos(\theta_2) \right]^T e^{j\omega t} e^{-j|\mathbf{R}||\mathbf{K}_2|\cos(\theta_2)}$$

$$\Rightarrow \mathbf{H}_{K2} = \frac{|\mathbf{E}|}{\eta_2} \left[0\ 1\ 0 \right]^T e^{j\omega t} e^{-j|\mathbf{R}||\mathbf{K}_2|\cos(\theta_2)}$$

Note:
$Hy \neq 0$ @z=0

TE Plane Wave at Oblique Incidence to Conductor

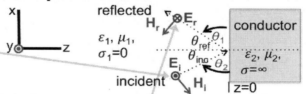

"TE polarization" because electric field is transverse to plane of incidence

- As shown, a conductor with boundary in the x-y plane, with incident $\mathbf{E}_i$ out of the page, and reflected $\mathbf{E}_r$ into the page

- The plane of incidence (the page) contains the normal to the boundary and the two plane wave wavevectors $\mathbf{K}_1$ and $\mathbf{K}_2$

- No electric or magnetic field is inside the conductor

- Same as prior slide, with $\mathbf{K}_2$ incident, $\mathbf{K}_1$ reflected wavevectors

- At the boundary z=0, $E_{t1} = 0$,
$\mathbf{E}_{10y} + \mathbf{E}_{20y} = 0$ at $z = 0$, for all x

Note minus sign

Note:
$H \neq 0$ @z=0

$$\Rightarrow 0 = |\mathbf{E}_{10y}| \left[0\ 1\ 0 \right]^T e^{j\omega t} e^{-j|\mathbf{x}|\beta_1 \cos(\theta_1)} + |\mathbf{E}_{20y}| \left[0\ -1\ 0 \right]^T e^{j\omega t} e^{-j|\mathbf{x}|\beta_2 \cos(\theta_2)}$$

$$\Rightarrow |\mathbf{E}_{10y}| = |\mathbf{E}_{20y}| \quad and \quad \beta_1 = \beta_2 \quad and \quad \theta_1 = \theta_2 \Rightarrow \theta_{inc} = \theta_{ref}$$

Plane Wave at Oblique Incidence to Dielectrics

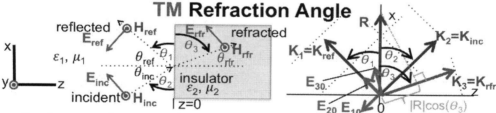

TM Refraction Angle

- As shown, a plane wave is incident on an insulator with **H** out of the page, where $\mathbf{K}_2$ is incident, $\mathbf{K}_1$ is reflected, $\mathbf{K}_3$ is refracted wavevector

- Tangential E-fields must be equal at boundary, at z=0, $E_{t1}=E_{t2}$, so:

at boundary: $E_{30}\left[\sin(\theta_3)\ 0\ -\cos(\theta_3)\right]^T e^{j\omega t}e^{-j|\mathbf{R}||\mathbf{K_3}|\cos(\theta_3)}$

$$= E_{10}\left[-\sin(\theta_1)\ 0\ -\cos(\theta_1)\right]^T e^{j\omega t}e^{-j|\mathbf{R}||\mathbf{K_1}|\cos(\theta_1)} + E_{20}\left[\sin(\theta_2)\ 0\ -\cos(\theta_2)\right]^T e^{j\omega t}e^{-j|\mathbf{R}||\mathbf{K_2}|\cos(\theta_2)}$$

tangential E-fields must be equal at z=0: so $\mathbf{E}_{K1x}+\mathbf{E}_{K2x}=\mathbf{E}_{K3x}$

$$\Rightarrow E_{20}\sin(\theta_2)e^{-j|\mathbf{R}|\beta_2\cos(\theta_2)} - E_{10}\sin(\theta_1)e^{-j|\mathbf{R}|\beta_1\cos(\theta_1)} = E_{30}\sin(\theta_3)e^{-j|\mathbf{R}|\beta_3\cos(\theta_3)}$$

$$\Rightarrow E_{inc}\cos(\theta_{inc})e^{-j|\mathbf{R}|\beta_2\cos(\theta_2)} - E_{ref}\cos(\theta_{ref})e^{-j|\mathbf{R}|\beta_1\cos(\theta_1)} = E_{rfr}\cos(\theta_{rfr})e^{-j|\mathbf{R}|\beta_3\cos(\theta_3)}$$

since spatial frequencies must be equal, then for $\beta_1=\beta_2$ then $\theta_1=\theta_2 \Rightarrow \boxed{\theta_{inc}=\theta_{ref}}$

then : $\beta_1\cos(\theta_1)=\beta_3\cos(\theta_3) \boxed{\Rightarrow \beta_1\sin(\theta_{inc})=\beta_3\sin(\theta_{rfr})}$ since $\cos(\pi/2-\phi)=\sin(\varphi)$

Snell's Law of Refraction

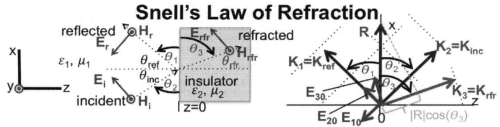

- As on the prior slide, a plane wave is incident on an insulator with **H** out of the page, where **K₂** is incident, **K₁** is reflected, and **K₃** is refracted wavevector

- For tangential E-fields equal at boundary, we saw:

prior results for $\beta_1 = \beta_2$:

$$\theta_{inc} = \theta_{ref} \quad and \quad \beta_1 \sin(\theta_{inc}) = \beta_3 \sin(\theta_{rfr})$$

- This can be rearranged into Snell's law of refraction:

refractive index $n = \sqrt{\mu_r \varepsilon_r}$, so: $\beta = 2\pi / \lambda = \omega / v_p = \omega\sqrt{\mu\varepsilon} = \omega\sqrt{\mu_r \varepsilon_r} / c = \omega n / c$

$$\theta_{inc} = \theta_{ref} \quad and \quad (\omega n_1 / c)\sin(\theta_{inc}) = (\omega n_2 / c)\sin(\theta_{rfr})$$

resulting in Snell's Law of refraction: $n_1 \sin(\theta_{inc}) = n_2 \sin(\theta_{rfr})$

TM Refraction Magnitudes

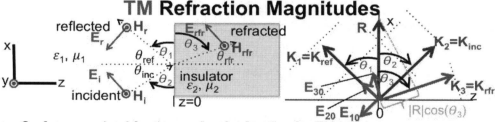

- So far, we solved for the angle of refraction for TM incidence
- Solution for the magnitudes follows similar lines, but details are omitted here
- Again, a plane wave is incident on an insulator with **H** out of the page, where **K₂** is incident, **K₁** is reflected, **K₃** is refracted wavevector

$$for : E_{inc} \cos(\theta_{inc}) e^{-j|R|\beta_2 \cos(\theta_2)} - E_{ref} \cos(\theta_{ref}) e^{-j|R|\beta_1 \cos(\theta_1)} = E_{rfr} \cos(\theta_{rfr}) e^{-j|R|\beta_3 \cos(\theta_3)}$$

Note: The direction of the reflection vector **E₂₀** can flip when Γ is negative

reflection: $\Gamma_{TM} = \dfrac{E_{ref}}{E_{inc}} = \dfrac{\eta_1 \cos(\theta_{inc}) - \eta_2 \cos(\theta_{rfr})}{\eta_2 \cos(\theta_{rfr}) + \eta_1 \cos(\theta_{inc})}$

transmission(refracted): $1 + \Gamma_{TM} = \dfrac{E_{rfr}}{E_{inc}} = \dfrac{2\eta_2 \cos(\theta_{inc})}{\eta_2 \cos(\theta_{rfr}) + \eta_1 \cos(\theta_{inc})}$

$where : \theta_{inc} = \theta_{ref}, \; n_1 \sin(\theta_{inc}) = n_2 \sin(\theta_{rfr}), \eta = \sqrt{\mu / \varepsilon}, \beta_1 = \beta_2, and \; n = \sqrt{\mu_r \varepsilon_r}$

TE Refraction Magnitudes and Angle

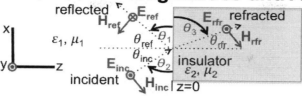

- Solution for TE refraction follows similar lines, but details are omitted here
- Again, a plane wave is incident on an insulator with **H** out of the page, where **K₂** is incident, **K₁** is reflected, **K₃** is refracted wavevector
- Solution:

$$for : E_{inc} \hat{\mathbf{y}} e^{-j|\mathbf{R}|\beta_2 \cos(\theta_2)} + E_{ref} \hat{\mathbf{y}} e^{-j|\mathbf{R}|\beta_1 \cos(\theta_1)} = E_{rfr} \hat{\mathbf{y}} e^{-j|\mathbf{R}|\beta_3 \cos(\theta_3)}$$

$$reflection: \quad \Gamma_{TE} = \frac{E_{ref}}{E_{inc}} = \frac{\eta_2 \cos(\theta_{inc}) - \eta_1 \sec(\theta_{rfr})}{\eta_1 \sec(\theta_{rfr}) + \eta_2 \sec(\theta_{inc})}$$

$$transmission(refracted): \quad 1 + \Gamma_{TM} = \frac{E_{rfr}}{E_{inc}} = \frac{2\eta_2 \sec(\theta_{inc})}{\eta_1 \sec(\theta_{rfr}) + \eta_2 \sec(\theta_{inc})}$$

$$where : \theta_{inc} = \theta_{ref}, \; n_1 \sin(\theta_{inc}) = n_2 \sin(\theta_{rfr}), \eta = \sqrt{\mu/\varepsilon}, \beta_1 = \beta_2, and \; n = \sqrt{\mu_r \varepsilon_r}$$

Total Internal Reflection

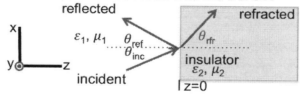

- When refractive index $n_1 > n_2$, it is possible to have total internal reflection when a plane wave is incident on a boundary
- The situation of total internal reflection occurs when:

$$\text{Total internal reflection condition: } \sin(\theta_{inc}) \geq \frac{n_2}{n_1} = \sqrt{\frac{\mu_{r2} \varepsilon_{r2}}{\mu_{r1} \varepsilon_{r1}}}$$

- The critical angle θ_c is the angle beyond which total internal reflection occurs :

$$\text{Total internal reflection critical angle: } \theta_c = \arcsin\left(\frac{n_2}{n_1}\right) = \arcsin\left(\sqrt{\frac{\mu_{r2} \varepsilon_{r2}}{\mu_{r1} \varepsilon_{r1}}}\right)$$

- Total internal reflection is the basic principle of optical fiber

8 WAVEGUIDE AND MICROSTRIP

This chapter covers a variety of methods for design and analysis of waveguide and microstrip

.

\

Coaxial Lines

TEM: **Lossless Coaxial Line**

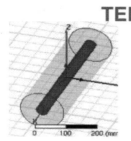 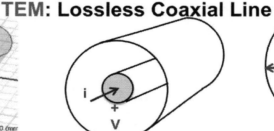 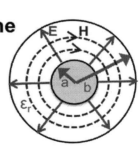

- A coaxial line consists of a metal center conductor of radius a and an outer conductor of radius b, with a dielectric of relative permittivity ε_r between
- With current i and voltage v as shown in the center figure above, the E-fields and H-fields would be as illustrated
- The fields are clearly TEM (transverse electromagnetic), with the direction of propagation, phase velocity, group velocity, and Poynting vector into the page

TEM: Lossless LC Model, C_R

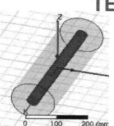
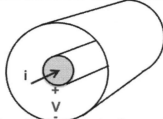
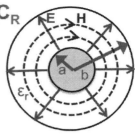

- The LC transmission line model of coaxial line follows from the theoretical inductance and capacitance of the coaxial line
- Let the center conductor be a cylindrical conductor sheet
- For a charge of ρ_L=1 C/m on a section of inner conductor of length l>>a, the E-field, voltage, and capacitance are

$$\mathbf{D} \approx \Psi / (2\pi rl) = \rho_L l / (2\pi rl) = \rho_L / (2\pi r), \text{ so } \mathbf{E} \approx \mathbf{D} / \varepsilon = \rho_L / (2\pi r\varepsilon)$$

$$V = \int_a^b \rho_L / (2\pi r\varepsilon) = \frac{\rho_L \ln(b/a)}{2\pi\varepsilon}, \text{ so } C = \frac{Q}{V} = \frac{\rho_L l(2\pi\varepsilon)}{\rho_L \ln(b/a)} = \frac{2\pi\varepsilon l}{\ln(b/a)} \text{ F}$$

- So C_R is

$$C_R = \frac{2\pi\varepsilon}{\ln(b/a)} \text{ F/m}$$

TEM: Lossless LC Model, L_R

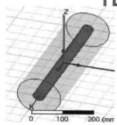
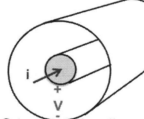
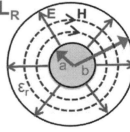

- Next, find L_R for LC transmision line approximation of coax
- Let the center conductor be a cylindrical current sheet
- For a current of I amperes on a section of inner conductor of length l>>a, the H-field, flux, and inductance are

$$\mathbf{H} \approx i / (2\pi r), \text{ so } \Phi = \int_S \mathbf{B}\,d\mathbf{S} \approx \int_a^b \frac{i\mu}{2\pi r} l\, dr = \frac{i\mu l \ln(b/a)}{2\pi}$$

$$\text{so } L = \frac{\Phi}{i} = \frac{\mu l \ln(b/a)}{2\pi} \text{ H}$$

- So L_R is

$$L_R = \frac{\mu \ln(b/a)}{2\pi} \text{ H/m}$$

TEM: Lossless Coaxial Line, Z_0, v_p

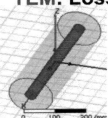 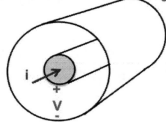 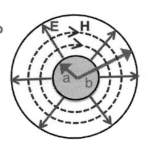

- Using the LC transmission line equations, and the foregoing results for C_R and L_R, then for a coaxial line, Z_0 and v_p are :

$$Z_0 = \sqrt{\frac{L_R}{C_R}} = \sqrt{\frac{\mu \ln(b/a)/2\pi}{2\pi\varepsilon/\ln(b/a)}} = \frac{\sqrt{\mu/\varepsilon}\,\ln(b/a)}{2\pi}$$

$$and \quad v_p = \frac{\omega}{\beta} = \lambda f = \frac{c}{\sqrt{\mu_r \varepsilon_r}}; \; \mu = \mu_r \mu_0, \varepsilon = \varepsilon_r \varepsilon_0$$

vacuum: $\mu = \mu_0 = 1257$ nH/m, $\quad \varepsilon = \varepsilon_0 = 8.85$ pF/m, $\quad c = 3\times10^8$ m/s

Parallel-Plate Lines

TEM: Parallel-Plate Transmission Line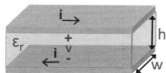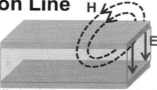

- A parallel-plate transmission line (or parallel-plate waveguide) consists of 2 metal strips of width w on a dielectric substrate of thickness h with relative permittivity ε_r
- With current i and voltage v as illustrated on the left, the E-fields and H-fields would appear as illustrated on the right
- The fields shown above are "quasi-TEM" since they are nearly TEM (transverse electromagnetic) for h<<w
- With h<<w the vast majority of fields lie within the dielectric substrate, such that the small amount of fringing fields near the edges can largely be ignored
- Later, we will see that parallel-plate transmission lines support other modes: TE (transverse electric) and TM (transverse magnetic)

TEM: Parallel-Plate Transmission Line

 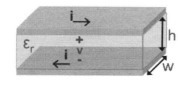 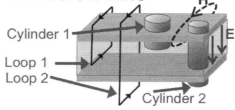

- **Analysis for h<<w**
- For current in the top plate equal to the bottom, in the integral of loop 2 above, ∮HdL=i-i=0, so H≈0 outside the transmission line
- For loop 1, ∮HdL=i, so H≈i/w is constant for any x coordinate
- In the integral of cylinder 2 above, ∮DdS = ∮εEdS =Q=0 so E≈0 outside the transmission line
- For cylinder 1, ∮εEdS =Q, so E is constant for any x coordinate
- These conditions (constant E and H) are satisfied by a plane wave $\mathbf{E}_x\, e^{-\gamma z} \Rightarrow \begin{bmatrix} E_{x0}\,0\,0 \end{bmatrix}^T e^{-\gamma z} e^{j\omega t}$ $where: \gamma = \pm j\omega\sqrt{\mu\varepsilon}$

$$\mathbf{H}_y e^{-\gamma z} \Rightarrow \frac{1}{\eta}\begin{bmatrix} 0\,E_{x0}\,0 \end{bmatrix}^T e^{-\gamma z} e^{j\omega t} \quad where: \quad \eta = \frac{j\omega\mu}{\gamma} = \sqrt{\frac{\mu}{\varepsilon}}$$

TEM: **Parallel-Plate Transmission Line**

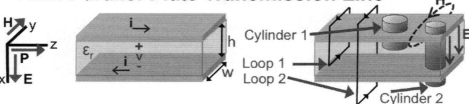

- Since H≈i/w we can calculate voltage v between plates

$$\mathbf{E}_x \, e^{-\gamma z} \Rightarrow \eta \begin{bmatrix} H_{y0} & 0 & 0 \end{bmatrix}^T e^{-\gamma z} e^{j\omega t} \quad where: \; \eta = \sqrt{\mu / \varepsilon}$$

$$so: v = \int_L \mathbf{E} \, d\mathbf{L} = \int_L \mathbf{E}_x \, d\mathbf{x} = \eta H_{y0} h = \sqrt{\mu / \varepsilon} \left(i / w \right) h$$

- Finally, we can calculate the characteristic impedance of the parallel-plate transmission line

$$Z_0 = v / i = \left(\sqrt{\mu / \varepsilon} \left(i / w \right) h \right) / i$$

$$= \left(h / w \right) \sqrt{\mu / \varepsilon} = 377 \left(h / w \right) \sqrt{\mu_r / \varepsilon_r} \;\; \text{ohms}$$

TEM: **Parallel-Plate Transmission Line**, LC Model

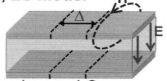

- The parallel plate line can also be analyzed as an LC transmission line, in terms of L_R and C_R
- Inductance for a section of length Δ and H=i/w yields

$$L = \frac{\Phi}{i} = \frac{\int_S \mathbf{B} \, d\mathbf{S}}{i} = \frac{\mu \left(i / w \right) h \Delta}{i} = \frac{\mu h \Delta}{w} \;\; \text{H, so} \;\; L_R = \frac{\mu h}{w} \;\; \text{H/m},$$

- Capacitance for a section of length Δ yields

$$C = \frac{\varepsilon w \Delta}{h} \;\; \text{F, so} \;\; C_R = \frac{\varepsilon w}{h} \;\; \text{F/m}$$

- For the LC transmission line model, the intrinsic impedance is

$$Z_0 = \sqrt{L_R / C_R} = \sqrt{\frac{\mu h / w}{\varepsilon w / h}} = \left(h / w \right) \sqrt{\mu / \varepsilon} \;\; \text{ohms, as before}$$

Parallel-Plate Waveguide Modes

Recall: TE Oblique Incidence to Conductor

"TE polarization" because electric field is transverse to plane of incidence

reflected

conductor

$\varepsilon_1, \mu_1,$
$\sigma_1 = 0$

θ_{ref} θ_1

θ_{inc} θ_2

$\varepsilon_2, \mu_2,$
$\sigma = \infty$

incident

z=0

- As shown, a conductor with boundary in the x-y plane, with incident E_i out of the page, and reflected E_r into the page

- The plane of incidence (the page) contains the normal to the boundary and the two plane wave wavevectors K_1 and K_2

- No electric or magnetic field is inside the conductor

- Same as prior slide, with K_2 incident, K_1 reflected wavevectors

- At the boundary z=0, $E_{t1}=0$,
  $\mathbf{E_{10y}} + \mathbf{E_{20y}} = 0$ at $z = 0$, for all x

Note minus sign

Note: $H \neq 0$ @z=0

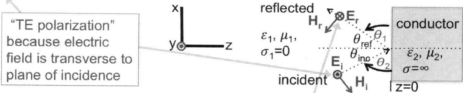

$$\Rightarrow 0 = \left|\mathbf{E_{10y}}\right| \begin{bmatrix} 0 & 1 & 0 \end{bmatrix}^T e^{j\omega t} e^{-j|\mathbf{x}|\beta_1 \cos(\theta_1)} + \left|\mathbf{E_{20y}}\right| \begin{bmatrix} 0 & -1 & 0 \end{bmatrix}^T e^{j\omega t} e^{-j|\mathbf{x}|\beta_2 \cos(\theta_2)}$$

$$\Rightarrow \left|\mathbf{E_{10y}}\right| = \left|\mathbf{E_{20y}}\right| \text{ and } \beta_1 = \beta_2 \text{ and } \theta_1 = \theta_2 \Rightarrow \theta_{inc} = \theta_{ref}$$

TE: Parallel-Plate Waveguide Cutoff

 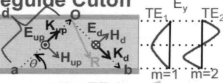

- A parallel-plate waveguide can support the TE (transverse electric) mode above, with E-field parallel to the plates, and wavevectors $\mathbf{K_{up}}=\beta[\sin(\theta)\ 0\ \cos(\theta)]$, $\mathbf{K_d}=\beta[-\sin(\theta)\ 0\ \cos(\theta)]$

- Recall, conductor zero-tangential E-field conditions above:

$$0=\left|\mathbf{E_{up}}\right|\begin{bmatrix}0 & 1 & 0\end{bmatrix}^T e^{j\omega t}e^{-j\mathbf{K_{up}}\cdot\mathbf{R}}+\left|\mathbf{E_d}\right|\begin{bmatrix}0 & -1 & 0\end{bmatrix}^T e^{j\omega t}e^{-j\mathbf{K_d}\cdot\mathbf{R}}$$

- Note, zero-tangential solution is periodic along x-axis above

$$0=\left|\mathbf{E_{upy}}\right|e^{j\omega t}e^{j\beta x\sin(\theta)}-\left|\mathbf{E_{dy}}\right|e^{j\omega t}e^{-jx\beta\sin(\theta)}$$

Note: mode integer m indicates m half-cycles of tangential E-field nulls

$$=j2\left|\mathbf{E}\right|e^{j\omega t}\sin\left(\beta x\sin(\theta)\right)=0\ for\ \beta x\sin(\theta)=m\pi,$$

$$\Rightarrow h\beta\sin(\theta)=h(2\pi/\lambda)\sin(\theta)=m\pi,\quad so\ h=m\lambda/\left(2\sin(\theta)\right)$$

$$or\ \ h\beta\sin(\theta)=h\left(2\pi f/v_p\right)\sin(\theta)=m\pi,\quad so\ f=mv_p/\left(2h\sin(\theta)\right)$$

cutoff frequency (minimum frequency) is $f_c=\dfrac{mv_p}{2h}=\dfrac{mc}{2h\sqrt{\varepsilon_r\mu_r}}$

TE: Parallel-Plate Waveguide β_{zm}

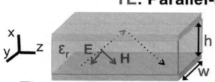 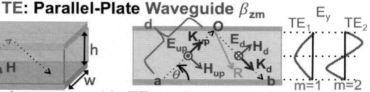

- The parallel-plate waveguide TE mode propagation constant β_z is taken from the z-component of the solution wavevectors $\mathbf{K_{up}}=\beta[\sin(\theta)\ 0\ \cos(\theta)]$, $\mathbf{K_d}=\beta[-\sin(\theta)\ 0\ \cos(\theta)]$, using the results from the previous slide

$$\cos(\theta)=\sqrt{1-\sin^2(\theta)}$$

$$\mathbf{E_{tot}}=\left|\mathbf{E_{up}}\right|\begin{bmatrix}0 & 1 & 0\end{bmatrix}^T e^{j\omega t}e^{-j\mathbf{K_{up}}\cdot\mathbf{R}}+\left|\mathbf{E_d}\right|\begin{bmatrix}0 & -1 & 0\end{bmatrix}^T e^{j\omega t}e^{-j\mathbf{K_d}\cdot\mathbf{R}}$$

$$=\left|\mathbf{E}\right|e^{j\omega t}\left(e^{j\beta x\sin(\theta)-j\beta z\cos(\theta)}-e^{-j\beta x\sin(\theta)-j\beta z\cos(\theta)}\right)=\left|\mathbf{E}\right|e^{j\omega t}e^{-j\beta z\cos(\theta)}\left(e^{j\beta x\sin(\theta)}-e^{-j\beta x\sin(\theta)}\right)$$

$$=j2\left|\mathbf{E}\right|e^{j\omega t}\sin\left(\beta x\sin(\theta)\right)e^{-j\beta z\cos(\theta)}=j2\left|\mathbf{E}\right|e^{j\omega t}\sin\left(xm\pi/h\right)e^{-j\beta z\cos(\theta)}$$

$$=j2\left|\mathbf{E}\right|e^{j\omega t}\sin\left(xm\pi/h\right)e^{-jz\sqrt{\beta^2-(m\pi/h)^2}}=j2\left|\mathbf{E}\right|e^{j\omega t}\sin\left(xm\pi/h\right)e^{-jz\beta\sqrt{1-(m\pi/(\beta h))^2}}$$

$$=j2\left|\mathbf{E}\right|e^{j\omega t}\sin\left(xm\pi/h\right)e^{-jz\beta\sqrt{1-(f_{cm}/f)^2}}=j2\left|\mathbf{E}\right|e^{j\omega t}\sin\left(xm\pi/h\right)e^{-j\beta_{zm}z}$$

$\mathrm{TE_m}$ waveguide "mode m" propagation constant

where $\beta_{zm}=\omega\sqrt{\mu\varepsilon}\sqrt{1-\left(f_{cm}/f\right)^2}$ and $f_{cm}=m/\left(2h\sqrt{\varepsilon\mu}\right)$ for integer m

TE: Parallel-Plate Maxwell Equations

 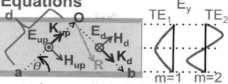

- The remaining TE field components in parallel-plate waveguide can be found from Maxwell's equations for source-free regions

$$\nabla \times \mathbf{E} = -j\omega\,\mathbf{B} = -j\omega\mu\,\mathbf{H} \qquad \nabla \times \mathbf{H} = \mathbf{J}_e + j\omega\,\mathbf{D} = j\omega\varepsilon\,\mathbf{E} \quad \text{source-free}$$

- Expanding for E-field and H-field:

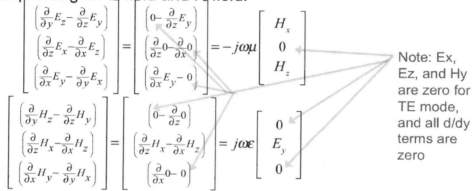

Note: Ex, Ez, and Hy are zero for TE mode, and all d/dy terms are zero

TM: Parallel-Plate Waveguide Cutoff

 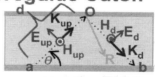

- A parallel-plate waveguide can support the TM (transverse magnetic) mode above, with H-field parallel to the plates, and wavevectors $\mathbf{K}_{up}=\beta[\sin(\theta)\ 0\ \cos(\theta)]$, $\mathbf{K}_d=\beta[-\sin(\theta)\ 0\ \cos(\theta)]$

$$\mathbf{E} = \left|\mathbf{E}_{up}\right|\left[\cos(\theta)\ 0\ -\sin(\theta)\right]^T e^{j\omega t}e^{-j\mathbf{K}_{up}\cdot\mathbf{R}} + \left|\mathbf{E}_d\right|\left[\cos(\theta)\ 0\ \sin(\theta)\right]^T e^{j\omega t}e^{-j\mathbf{K}_d\cdot\mathbf{R}}$$

- Note, zero-tangential solution is periodic along x-axis above

$$0 = \mathbf{E}_{upx} + \mathbf{E}_{dx} = -\left|\mathbf{E}_{up}\right|\sin(\theta)e^{j\omega t}e^{j\beta x\sin(\theta)} + \left|\mathbf{E}_d\right|\sin(\theta)e^{j\omega t}e^{-j\beta x\sin(\theta)}$$

Note: mode integer m indicates m half-cycles of tangential E-field nulls

$$= -j2\left|\mathbf{E}\right|e^{j\omega t}\sin(\theta)\sin\left(\beta x\sin(\theta)\right) = 0\ for\ \beta x\sin(\theta) = m\pi,$$

$$\Rightarrow h\beta\sin(\theta) = h\left(2\pi/\lambda\right)\sin(\theta) = m\pi, \quad so\ h = m\lambda/\left(2\sin(\theta)\right)$$

$$or\ \ h\beta\sin(\theta) = h\left(2\pi f/v_p\right)\sin(\theta) = m\pi, \quad so\ f = mv_p/\left(2h\sin(\theta)\right)$$

cutoff frequency (minimum frequency) is $f_c = \dfrac{mv_p}{2h} = \dfrac{mc}{2h\sqrt{\varepsilon_r\mu_r}}$

TM: Parallel-Plate Waveguide β_z

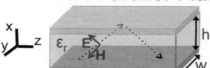
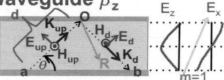

- Then, parallel-plate waveguide TM mode propagation constant β_z is taken from the z-component of the solution wavevectors

$$\mathbf{E}_{tot} = |\mathbf{E}_{up}| \left[\cos(\theta)\ 0\ -\sin(\theta) \right]^T e^{j\omega t} e^{-j\mathbf{K}_{up} \cdot \mathbf{R}} + |\mathbf{E}_d| \left[\cos(\theta)\ 0\ \sin(\theta) \right]^T e^{j\omega t} e^{-j\mathbf{K}_d \cdot \mathbf{R}}$$

$$= |\mathbf{E}| \begin{bmatrix} \cos(\theta)e^{j\omega t}\left(e^{j\beta x\sin(\theta)-j\beta z\cos(\theta)} + e^{-j\beta x\sin(\theta)-j\beta z\cos(\theta)} \right) \\ 0 \\ -\sin(\theta)e^{j\omega t}\left(e^{j\beta x\sin(\theta)-j\beta z\cos(\theta)} - e^{-j\beta x\sin(\theta)-j\beta z\cos(\theta)} \right) \end{bmatrix}$$

Note: E_x, is not zero near the plates, but tangential E_z is zero near plates

$$= |\mathbf{E}| e^{j\omega t} e^{-j\beta z\cos(\theta)} \begin{bmatrix} 2\cos(\theta)\cos(\beta x\sin(\theta)) \\ 0 \\ 2j\sin(\theta)\sin(\beta x\sin(\theta)) \end{bmatrix} = |\mathbf{E}| e^{j\omega t} e^{-j\beta_{zm} z} \begin{bmatrix} 2\cos(\theta)\cos(\beta x\sin(\theta)) \\ 0 \\ 2j\sin(\theta)\sin(\beta x\sin(\theta)) \end{bmatrix}$$

where $\beta_{zm} = \omega\sqrt{\mu\varepsilon}\sqrt{1-\left(f_{cm}/f\right)^2}$ and $f_{cm} = m/\left(2h\sqrt{\varepsilon\mu}\right)$ for integer m

TM: Parallel-Plate Maxwell Equations

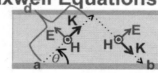

- The remaining TM field components in parallel-plate waveguide can be found from Maxwell's equations for source-free regions

$$\nabla \times \mathbf{E} = -j\omega\,\mathbf{B} = -j\omega\mu\,\mathbf{H} \qquad \nabla \times \mathbf{H} = \mathbf{J}_e + j\omega\,\mathbf{D} = j\omega\varepsilon\,\mathbf{E} \quad \text{source-free}$$

- Expanding for E-field and H-field:

$$\begin{bmatrix} \left(\frac{\partial}{\partial y}E_z - \frac{\partial}{\partial z}E_y\right) \\ \left(\frac{\partial}{\partial z}E_x - \frac{\partial}{\partial x}E_z\right) \\ \left(\frac{\partial}{\partial x}E_y - \frac{\partial}{\partial y}E_x\right) \end{bmatrix} = \begin{bmatrix} \left(0-\frac{\partial}{\partial z}0\right) \\ \left(\frac{\partial}{\partial z}E_x - \frac{\partial}{\partial x}E_z\right) \\ \left(\frac{\partial}{\partial x}0-0\right) \end{bmatrix} = -j\omega\mu \begin{bmatrix} 0 \\ H_y \\ 0 \end{bmatrix}$$

$$\begin{bmatrix} \left(\frac{\partial}{\partial y}H_z - \frac{\partial}{\partial z}H_y\right) \\ \left(\frac{\partial}{\partial z}H_x - \frac{\partial}{\partial x}H_z\right) \\ \left(\frac{\partial}{\partial x}H_y - \frac{\partial}{\partial y}H_x\right) \end{bmatrix} = \begin{bmatrix} \left(0-\frac{\partial}{\partial z}H_y\right) \\ \left(\frac{\partial}{\partial z}0-\frac{\partial}{\partial x}0\right) \\ \left(\frac{\partial}{\partial x}H_y - 0\right) \end{bmatrix} = j\omega\varepsilon \begin{bmatrix} E_x \\ 0 \\ E_z \end{bmatrix}$$

Note: Ey, Hx, and Hz are zero for TM mode, and all d/dy terms are zero

138

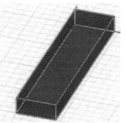

Rectangular Waveguide

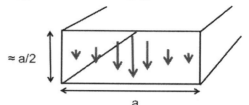

$\approx a/2$

a

- Rectangular waveguide
- Frequencies below cutoff cannot propagate
  - Cutoff = $1.5 \times 10^8/a = c/(2a)$
    - where a= longer dimension in meters
    - Smaller dimension is usually approximately a/2
  - WR42: 10.7x4.3mm; 14GHz cutoff (18-26.5 band)
- TE_{10} mode, transverse electric field across short dimension
- Only TE_{10} mode propagates from cutoff to nearly twice cutoff, and the recommended upper operating frequency is less than this

Cavity Resonators

- A cavity resonator can be thought of as the electromagnetic equivalent of an LC inductor-capacitor resonator
- Ideally, the resonator is lossless
- A simple electromagnetic version of a resonator could be a coaxial line with short circuits at both ends. If the line was lossless, any wave would endlessly reflect from both ends
- The coaxial resonator would support sinusoidal standing waves of multiples of a half wavelength
- Similarly, other waveguide structures can support standing waves and form resonator cavities
- High-Q resonators have lower loss, just as for parallel RLC circuits where $Q = R_p/(\omega_0 L_p) = \omega_0 C_p R_p$ and $\omega_0 = 1/(L_p C_p)^{1/2}$

Microstrip Lines

Microstrip

 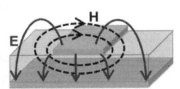

- A microstrip line consists of a metal strip of width w on a dielectric substrate of thickness h with relative permittivity ε_r
- With current i and voltage v as illustrated on the left, the E-fields and H-fields would appear as illustrated on the right
- The fields are "quasi-TEM" since they are nearly TEM for h<<w
- Because the phase velocity of the fields in air does not equal the phase velocity in the dielectric substrate, it is impossible to have "pure TEM" fields with solutions for the phases of the fields in air and substrate matched in a perfect TEM solution
- Nevertheless, the ease of fabrication on pc boards makes microstrip one of the most useful transmission lines

From: E. O. Hammerstad, "Equations for Microstrip Circuit Design," 1975 5th European Microwave Conference, Hamburg, Germany, 1975, pp. 268-272. and cited by I. J. Bahl and D. K. Trivedi, "A Designer's Guide to Microstrip Line," Microwaves, p. 174, May, 1977

Microstrip Design Formulas: w/h<1

- For w/h<1
- Cutoff frequency$=106/[h(\varepsilon_r)^{1/2}]$ GHz, for h in mm
- For microstrip line with width w, dielectric height h:
  $$Z_0 \approx \{ 60/(\varepsilon_e)^{1/2} \} \ln(8\,h/w + 0.25\,w/h)$$
  Velocity v in meter/second
  $$v = c / (\varepsilon_e)^{1/2}$$
  Effective dielectric constant
  $$\varepsilon_e = (\varepsilon_r+1)/2 + (\varepsilon_r-1) [(1 + 12\,h/w)^{-1/2} +0.04(1-w/h)^2]/2$$
- Also: microstrip is dispersive, but we will not cover details

From: E. O. Hammerstad, "Equations for Microstrip Circuit Design," 1975 5th European Microwave Conference, Hamburg, Germany, 1975, pp. 268-272. and cited by I. J. Bahl and D. K. Trivedi, "A Designer's Guide to Microstrip Line," Microwaves, p. 174, May, 1977

Microstrip Design Formulas: w/h>1

- For w/h>1
- Cutoff frequency$=106/[h(\varepsilon_r)^{1/2}]$ GHz, for h in mm
- For microstrip line with width w, dielectric height h:
  $$Z_0 \approx 377 / [(\varepsilon_e)^{1/2} \{1.393 + w/h + 0.667 \ln(w/h + 1.44) \}]$$
  Velocity v in meter/second
  $$v = c / (\varepsilon_e)^{1/2}$$
  Effective dielectric constant
  $$\varepsilon_e = (\varepsilon_r+1)/2 + (\varepsilon_r-1) / [2 (1 + 12\,h/w)^{1/2}]$$
- Typical dielectrics:
  Polyethylene: $\varepsilon_r = 2.6$ FR4 (or G10) $\varepsilon_r = 4.3$
  Teflon $\varepsilon_r = 2.1$ Epsilam10 $\varepsilon_r = 10$

From: E. O. Hammerstad, "Equations for Microstrip Circuit Design," 1975 5th European Microwave Conference, Hamburg, Germany, 1975, pp. 268-272. and cited by I. J. Bahl and D. K. Trivedi, "A Designer's Guide to Microstrip Line," Microwaves, p. 174, May, 1977

Microstrip: Example

- Example 50 ohm microstrip line:
  FR-4: w/h ≈ 2 for a 50 ohm line

  $\varepsilon_e = (\varepsilon_r+1)/2 + (\varepsilon_r-1) + (\varepsilon_r-1) / [2(1 + 12\,h\,/\,w)^{1/2}]$

  $\quad = (4.3+1)/2 + (4.3-1) / [2(1 + 6\,)^{1/2}]$

  $\quad = 2.65 + .62 = 3.27$

  $Z_0 \approx 377 / [\,(\varepsilon_e)^{1/2}\,\{1.393 + w/h + 0.67\,\ln(w/h + 1.44)\,\}\,]$

  $\quad = 49.4\,\Omega$

  $v = c / (3.27)^{1/2} = 0.55\,c$

- Loss/m = 0.5 dB/inch = 20 dB/m @ 1 GHz
- Generally, the behavior of microstrip is very complex and changes with frequency. Sophisticated computer modeling is required at high frequency.

9 ANTENNAS, RADIATION, AND RADIO

This chapter provides discussion of the design and analysis of antennas, radiation, and radio systems.

Free Space Propagation

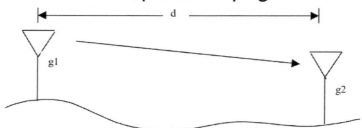

- 2 Antennas, with gains g_1 and g_2 (linear, not dB)
  - $G_1 = 10\log_{10}(g_1)$ is antenna gain in dB
- Separation d in meters
- Wavelength $\lambda = c/f$, c=speed of light in m/s
- P_r= received power, and P_t = transmitted power
- The free-space propagation loss as a linear ratio is

$$\frac{P_r}{P_t} = \frac{g_1 \cdot g_2 \cdot (\lambda)^2}{(4 \cdot \pi)^2 \cdot d^2}$$

289

Propagation Loss in dB

- Free Space loss in dB:
  - $L_P = -10 \log_{10}(P_r/P_t)$
- Antenna gains in dBi:
  - $G_1 = 10 \log_{10}(g_1)$
  - $G_2 = 10 \log_{10}(g_2)$

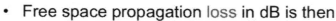

- Free space propagation loss in dB is then

  $L_p = -10 \log_{10}(\lambda^2) + 10 \log_{10}[(4\pi d)^2] - G_1 - G_2$

  where wavelength $\lambda = c/f$ m

  and distance d is in meters

- Loss increases 20 dB for 10 times distance
- Loss increases 6 dB for 2 times distance
- Loss increases 20 dB for 10 times frequency

Propagation Loss Example

- Find the free-space path loss at when:
  - Signal is at 100 MHz
  - distance between the antennas is 4 kilometers (2.5 mile)
  - gain of the transmitting antenna is 10 dBi,
  - gain of the receiving antenna is 2 dBi
- Then:
  - $\lambda = c / f = 3$ m
  - d=4000 m
  - $L_p = -10 \log_{10}(\lambda^2) + 10 \log_{10}[(4\pi d)^2] - G_1 - G_2$
    $= 84.5 - 10 - 2 = 72.5$ dB
- For this example, the receiver signal will be 72.5 dB below the transmitted signal power

Urban Propagation Loss Models

- The free-space model does not represent urban areas well
- So, other models are used
- A more general urban model (approx.) for propagation loss in dB would be of the form:

  $$L_p = -10 \log_{10}(\lambda^2) + 10 \log_{10}[(4\pi)^2 d^n] - G1 - G2$$

- Parameter n=2 for free space model
- Typical values of n:

  Free space, n = 2

  Urban areas, n = 2.7 - 5

  Indoors,   n = 2 - 6

- One of the most commonly used urban models is the Hata model, where the exponent is 2.6, but other factors are also added in the Hata formula

Fading

- Fading denotes random variations in radio signal strength
- It is caused by multipath, diffraction, etc.
- Mobile systems exhibit fast fading with motion
- Fade margin is an "extra signal power" design margin or allowance based on the percentage of time signal is to remain above average levels
- Rule of thumb:
  - Fade margin =10 dB for 90% of time
  - Fade margin =20 dB for 99% of time
  - Fade margin =30 dB for 99.9% of time
- So, increasing power 20 dB above the designed average minimum signal strength of a radio system would result in a system that "drops out" less than 1% of the time due to fading

Antenna Basics

- Antenna gain in dBi is dB gain relative to isotropic antenna
- Antenna gain in dBd is dBgain relative to a half-wave dipole antenna
- Gain in dBi ≈ gain in dBd + 2
- Higher gain results in more directional antenna with narrowed beamwidth
- Antenna patterns plot radiation strength as function of direction

Antenna Basics

- Effective area (or effective aperture)
$$A_e = g\,\lambda^2 / (4\pi)$$
  where g is gain (linear) of antenna, $g = 10^{G/10}$
- Beamwidth in degrees $\approx 229 / g^{0.5}$
- Antennas have input impedance
- Antenna Polarization: vertical, horizontal, circular
  (direction of E-field vector)
- Typically require balun to change a shielded or unbalanced feed to a balanced feed
- Antenna efficiency η = power actually transmitted by antenna divided by total power delivered to antenna
- Gain g= η d where d is directivity, and D=10 $\log_{10}$(d) dB

Half Wave Dipole

- "Rabbit Ears"
- Two quarter-wave elements
- Total length = $\lambda / 2$
- Input impedance: Z_{in} = 73 ohms typical
- Antenna gain ≈ 2 dBi
- Bandwidth ≈ 5 - 30 %
- Polarization: Linear (vertical or horizontal)

Monopole Antenna Above Ground Plane

- "Vertical whip"
- Total length (or height) = $\lambda/4$
- Input impedance: Zin = 37 ohms typical
- Antenna gain ≈ 2 dBi
- Bandwidth ≈ 5 - 30 %
- Polarization: Linear (vertical)
- The ground plane can be thought of as a mirror. where the "other half of the dipole" is the reflection of the monopole in the ground plane

Other Antennas

- Parabolic dish (high gain)
- Yagi
- Spiral (very wide bandwidth)
- Waveguide horn
- Phased Array
- Loop
- Folded dipole (300 ohm)
- Ferrite rod

Dish Antenna Example

- Estimate the gain and beamwidth of a 1 meter diameter parabolic antenna at 18 GHz.

  - λ = c/f = 0.017m
  - Effective area: $A_e = g \lambda^2 / (4 \pi) = \pi r^2 = 0.8m^2$
  - So: $g = 4 \pi A_e / \lambda^2 = 34{,}785$
  - And: $G = 10 \log_{10}(g)$
      = $10 \log_{10}[4 \pi 0.8 / (0.017)^2] = 45.4$ dB
  - Beamwidth in degrees $\approx 229 / g^{0.5} = 1.2$ degrees

Wheeler-Chu Limit

- Large antennas are incompatible with mobile devices
- Half-wave dipole ~1.5 meters at VHF (~100 MHz)
- Electrically small antennas have limited bandwidth
- Wheeler-Chu limit for antenna size fitting within radius a.
  - Bandwidth B in Hz decreases as $(ka)^3 = (2\pi a/\lambda)^3$
  - $Q \approx f_0/B \approx 1/(ka)^3$ for ka<<1 at center frequency f

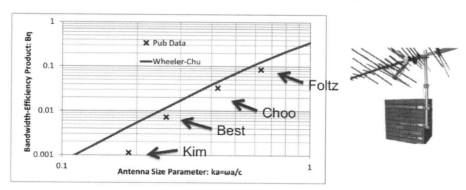

Potential Power Savings

- Large potential power savings moving from 1 GHz to 100 Hz
- Save up to 99% energy per decade
  - Friis formula
    - Path loss decreases 20 dB/decade
  - Hata formula
    - Path loss decreases 26 dB/decade
- Tradeoffs
  - Use non-Foster devices to recover bandwidth
    - Use negative capacitance and/or inductance
  - Small antennas have ≈ same gain (for same efficiency)
    - Efficiency issues may limit results
  - Non-Foster does require some power

Digital Non-Foster
Wideband Electrically-Small Antennas

Analog Non-Foster Circuits

- ## Of particular interest:
  - o Negative capacitors —‖—
  - o Negative inductors ⌒⌒⌒⌒

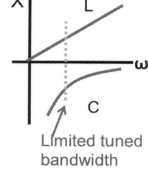

Limited tuned bandwidth

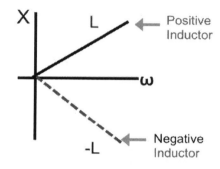

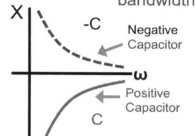

Digital Impedance Circuits

Simple Example : Digital Resistor

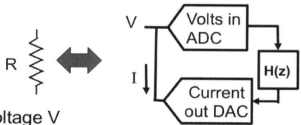

•Measure voltage V
•Set current I
•Let H(z) = 1/R
•So, output current: I = V/R

... yields world's most expensive resistor!
... but is tunable
... Useful in implementing "exotic" impedances

Digital Is The Key (Digital Non-Foster)

- Digital Non-Foster Circuits
  - Repeatability and accuracy (1's & 0's)
  - Can tightly control stability
  - Nyquist speed limit for stability
  - Can adapt to antenna impedance changes
  - Can implement whole circuits easily
  - Offers adaptability/repeatability in arrays
  - It does not have to be non-Foster

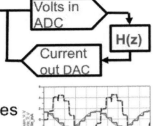

- Analog Non-Foster Circuits
  - Analog technology
  - Non-Foster notoriously unstable
  - Tolerance, temperature, reproducibility
  - Antenna impedance variations

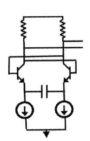

Theory: Digital Impedance

- Measure v(t)   ADC: $v[n] = v(nT)$
- Calculate $v_{dac}[n]$ H(z): $v_{dac}[n] = v[n]*h[n]$
- Set $v_{dac}(t)$ DAC: $v_{dac}(nT) = v_{dac}[n]$

$$V_{dac}(s) = V_{in}^{*}(s)H(z)e^{-s\tau}\left.\frac{(1-z^{-1})}{s}\right|_{z=e^{sT}} \qquad I_{in}(s) \approx \frac{V_{in}(s) - V_{dac}(s)}{R_{dac}}$$

$$V_{dac}(z) = H(z)V_{in}(z)$$

$$V^{*}(s) = \sum v(nT)e^{-nsT}$$
$$= \sum V(s - n\omega_0)/T$$
(Starred Transform)

Design H(z) to obtain desired impedance

Input impedance is then:

$$Z(s) = \frac{V_{in}(s)}{I_{in}(s)} \approx \left.\frac{sTR_{dac}}{\left[sT - H(z)\left(1-z^{-1}\right)e^{-s\tau}\right]}\right|_{z=e^{sT}}$$

See: Weldon et al., "Thevenin Forms of Digital Discrete-Time Non-Foster RC and RL Circuits," in 2016 IEEE Int. Symp on Antennas and Prop.

Theory: Alternative Digital Impedance Circuit

- Measure v(t)   ADC: $v[n] = v(nT)$
- Calculate $i_{dac}[n]$ H(z): $i_{dac}[n] = v[n]*h[n]$
- Set $i_{dac}(t)$ DAC: $i_{dac}(nT) = i_{dac}[n]$

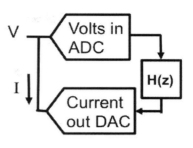

$$I_{dac}(s) = V_{in}^*(s)H(z)\frac{(1-z^{-1})}{s}\bigg|_{z=e^{sT}}$$

$$I_{dac}(z) = H(z)V_{in}(z)$$

$$V^*(s) = \sum v(nT)e^{-nsT}$$

$$= \sum V(s - n\omega_0)/T$$

(Starred Transform)

Input impedance is then:

$$Z(s) = \frac{V_{in}(s)}{I_{in}(s)} \approx \frac{sT}{(1-z^{-1})H(z)}\bigg|_{z=e^{sT}}$$

Design H(z) to obtain desired impedance

See: Weldon et al., "Performance of Digital Discrete-Time Implementations of Non-Foster Circuit Elements," in 2015 IEEE International Symposium on Circuits and Systems

Approach: Non-Foster RLC Match

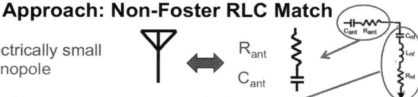

Electrically small monopole ⟷ R_{ant} C_{ant}

- Electrically-small monopole ≈ RC for ka<<1

Digital Non-Foster RLC matching network

C_{nf}
L_{nf}
R_{nf}

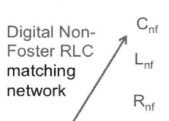

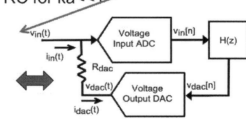

- Digital non-Foster RLC design approach, for example:
  - Use negative C_{nf} to cancel <u>portion</u> of C_{ant} and lower Q
  - Use positive L_{nf} to cancel remaining reactance
  - Use bilinear transform to find H(z) for RLC

See: Weldon et al., "Bilinear Transform Approach for Wideband Digital Non-Foster Matching of Electrically-Small Antennas," in 2018 IEEE Int. Symp on Antennas and Prop.,July 8-13, 2018

Approach: Example

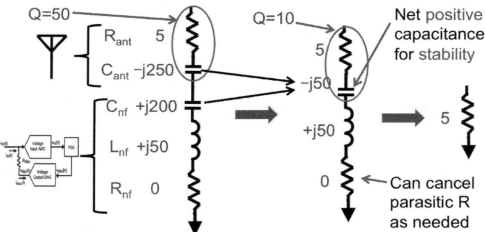

- Digital non-Foster RLC design approach:
  - Use negative C_{nf} to cancel portion of C_{ant} and lower Q
  - Use positive L_{nf} to cancel remaining reactance
  - Use bilinear transform to find H(z) for RLC

Simulation: Digital Non-Foster RLC Matched Monopole
- Comparison of passive vs. non-Foster matched antenna
  - 0.75 m ($\lambda/20$) monopole, $\approx$10x bandwidth improvement
  - 6 dB non-Foster bandwidth 19.0-23.8 MHz = 22%
  - Passive match 20.2-20.7MHz=2.4% vs. Chu 3.3%

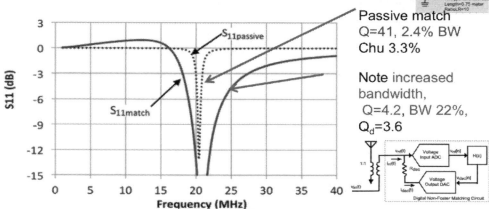

C_{nf}= -36.2 pF, L_{nf}=70 nH, R_{nf}=25, R_p=7,000, for T=1 ns, R_{dac}=200, Q_d=3.6

See: Weldon et al., "Bilinear Transform Approach for Wideband Digital Non-Foster Matching of Electrically-Small Antennas," in 2018 IEEE Int. Symp on Antennas and Prop.,July 8-13, 2018

Simulation: Digital Non-Foster RLC Matched Monopole

- Comparison of matched and unmatched antenna impedance
  - 0.75 m ($\lambda/20$) monopole, ka=0.31@20 MHz

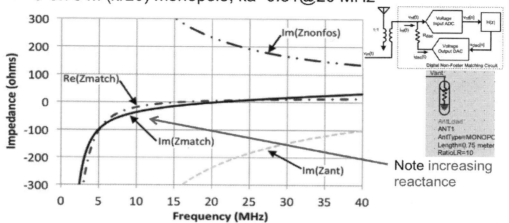

C_{nf}= -36.2 pF, L_{nf}=70 nH, R_{nf}=25 ohms, and R_p=7,000 ohms, for T=1 ns, R_{dac}=200 ohms, Q_d=3.6

3D Electromagnetic Simulation

3D Electromagnetic Simulation

- 3D Electromagnetic Simulation
  - Enter 3D physical structures
  - Enter material properties
  - Enter excitation ports
  - Simulator solves Maxwell equations
  - Outputs
    - S-parameters
    - Antenna patterns

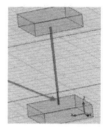

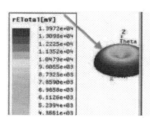

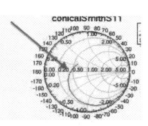

3D Electromagnetic Simulation of Antennas

- 3D Electromagnetic Simulation of antennas
  - Antenna structures can be entered
  - Antenna must have a signal input port
  - After simulation, s-parameters and antenna radiation patterns can be plotted

Antenna Structure   Input Impedance   Radiation Pattern

Radio Basics

Basic Superheterodyne Receiver

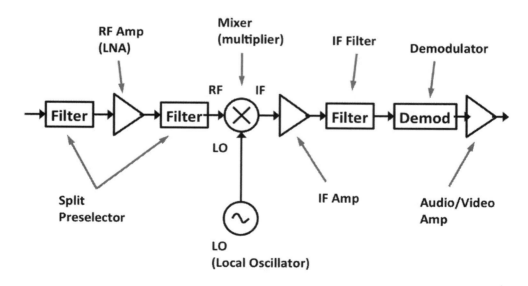

Radio Stages Summary

- Input from antenna
- Typically filtered by preselector before any amplifier stages
  - Eliminate spurs, image, out-of-band signals, radiation
- RF amplifier
  - Low noise, highest freq., large dyn. Range
  - Overcome mixer loss, LO radiation
- First mixer, first LO (tunable), first IF
- IF filter and amplifier
  - Adjacent channel selectivity
- Demodulator and amplifier
- Recover and amplify baseband signal

Digital Radio Introduction

- Digital radio includes a wide range of systems
  - SDR (software defined radio)
    - Ideally: just software+processor+ADC+DAC
  - Cognitive radio
    - A programmable digital radio with intelligent adaptability
  - Digital TV, digital cellphones, etc.
    - These categories sometimes refer to "digital data" being transmitted
- History and current status
  - According to Cruz et al. in *IEEE Microwave Mag.*, Jun 2010, SDR first appeared in Mitola in *IEEE Commun. Mag.*, May 1995
  - According to Cruz et al. in *IEEE Microwave Mag.*, Jun 2010, cognitive radio was proposed by Mitola & Maguire, *IEEE Pers. Comm.*, Aug. 1999
  - According to Hueber et al. in *IEEE Microwave Mag.*, Aug 2015, digital RF is now the predominant architecture in entry-level cell phones
  - According to Hueber et al. in *IEEE Microwave Mag.*, Aug 2015, 28-nm CMOS has 400 GHz f_T.
  - "Software radio" was also noted in May 1985 *E-systems Team*

Digital Radio Systems

- The "ideal digital radio" is simply an ADC (analog-to-digital converter), DAC (digital-to-analog converter), an antenna, and a DSP (digital signal processing) system
- The entire radio is implemented in signal processing
- Therefore, some key radio concepts will first be reviewed
- The concepts apply to a broad range of systems such as cellphones, WiFi, digital television, radar, and modems

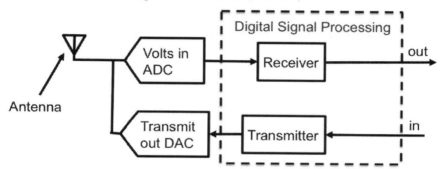

Quadrature Amplitude Modulation Theory

- Summary of theory of a QAM transmitter and receiver

$$f_{qam}[n] = m_c[n]\cos[\omega_c n] + m_s[n]\sin[\omega_c n]$$

$$\mathcal{F}\{m_c[n]\cos[\omega_c n]\} = 0.5\left[M_c(\omega - \omega_c) + M_c(\omega + \omega_c)\right]$$

$$\mathcal{F}\{m_s[n]\sin[\omega_c n]\} = -0.5j\left[M_s(\omega - \omega_c) - M_s(\omega + \omega_c)\right]$$

receiver cosine channel output :

$$\cos[\omega_c n]\left(m_c[n]\cos[\omega_c n] + m_s[n]\sin[\omega_c n]\right)$$

$$= 0.5\left\{m_c[n]\left(1 + \cos[2\omega_c n]\right) + m_s[n]\sin[2\omega_c n]\right\}$$

and after lowpass filter, receiver output $= m_c[n]$

receiver sine channel output :

$$\sin[\omega_c n]\left(m_c[n]\cos[\omega_c n] + m_s[n]\sin[\omega_c n]\right)$$

$$= 0.5\left\{m_c[n]\sin[2\omega_c n] + m_s[n]\left(1 - \cos[2\omega_c n]\right)\right\}$$

and after lowpass filter, receiver output $= m_s[n]$

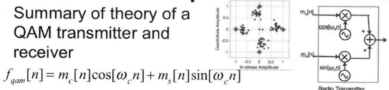

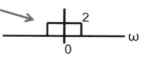

10 METAMATERIALS

This chapter covers the basics of metamaterials.

Metamaterials: Key People and Events

- 1968 Victor Veselago paper describing theoretical properties of metamaterials
- 1996 John Pendry et al. work on thin-wire negative dielectric
- 1999 John Pendry work on negative magnetic elements
- 2000 David R. Smith (now at Duke) research team at UCSD demonstrates first double-negative material
- 2006 Cloaking demonstrated by Schurig, Mock, Justice,. Cummer, Pendry, Starr, and Smith at Duke
- 2011 Gregoire, White, Colburn at HRL demonstrate wideband magnetic conductors

Veselago: Classes of Metamaterials

- Victor Veselago 1968 paper
- Normal materials: $\varepsilon > 0$ & $\mu > 0$
  - DPS (double-positive) (DPS) $\varepsilon > 0$ & $\mu > 0$
- Metamaterials
  - ENG (epsilon negative)
    - Negative permittivity, $\varepsilon < 0$
  - DNG (double negative)
    - Both negative, $\varepsilon < 0$ & $\mu < 0$
  - MNG (mu negative)
    - Negative permeability, $\mu < 0$

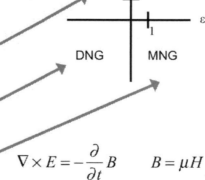

$$\nabla \times E = -\frac{\partial}{\partial t} B \qquad B = \mu H$$

$$\nabla \times H = \frac{\partial}{\partial t} D + J \qquad D = \varepsilon E$$

Capacitors & Inductors and ε and μ

- Consider electric field with conductors at equipotential planes
- A negative capacitor between plates can mimic a negative dielectric material

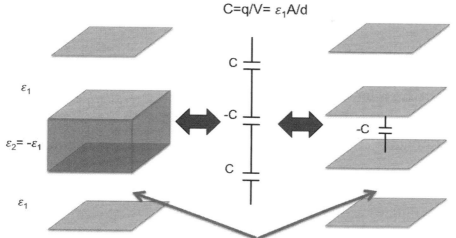

$$C = q/V = \varepsilon_1 A/d$$

Fields in these regions must be identical

Metamaterial Applications

1. Magnetic Conductors
   - Metal:
     - Charge moves freely
     - Has been around for 5000 years!
   - Wideband magnetic conductor:
     - Magnetic monopoles move freely
     - Gregoire, White, Colburn, 2011 (HRL/Boeing/General Motors)

Electric Conductor   Magnetic Conductor

2. Antennas
   - Improved performance
   - Reduced size
   - Next-generation cellphone/Wi-Fi

Non-Foster Applications

3. Cloaking
   - Invisibility cloaking
   - Schurig, et al. 2006

4. Fast-Wave Lines
   - Used in leaky-wave antennas
   - CRLH lines

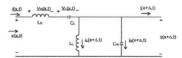

5. And more
   - Negative refraction
   - Lenses beyond diffraction limit
   - Carpet cloak
   - Acoustics
   - Mechanical

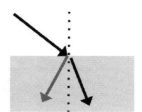

Metamaterial "Right-Handed" Transmission Lines

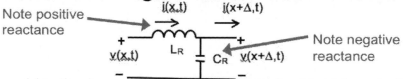

Note positive reactance

Note negative reactance

- Consider the lumped-element model of a "right-handed" transmission line shown above
- We will consider "left-handed" lines later
- The right-handed transmission line equations are

L is analogous to μ_R

$$\frac{\partial v(x,t)}{\partial x} = -L_R \frac{\partial i(x,t)}{\partial t} \quad \Rightarrow \quad \nabla \times E = -\mu_R \frac{\partial H}{\partial t}$$

$$\frac{\partial i(x,t)}{\partial x} = -C_R \frac{\partial v(x,t)}{\partial t} \quad \Rightarrow \quad \nabla \times H = \varepsilon_R \frac{\partial E}{\partial t}$$

C is analogous to ε_R

- The circuit model on the left corresponds to the three-dimensional Maxwell equations on the right
- Units are L_R in H/m and C_R in F/m

See: Weldon, et al., "`Left-Handed Extensions of Maxwell's Equations for Metamaterials," IEEE SoutheastCon 2010 Proceedings, March 18-21, 2010.

Propagation in "Right-Handed" Transmission Lines

$$\nabla \times E = -\mu_R \frac{\partial H}{\partial t}$$

$$\nabla \times H = \varepsilon_R \frac{\partial E}{\partial t}$$

- Taking the curl of both sides

$$\nabla \times \left(\nabla \times E = -\mu_R \frac{\partial H}{\partial t} \right) \Rightarrow \nabla \times \nabla \times E = -\mu_R \frac{\partial \nabla \times H}{\partial t}$$

- Then, the wave equation follows as (where $\nabla \cdot E = 0$):

$$\nabla \times \nabla \times E = \nabla \left(\nabla \cdot E \right) - \nabla^2 E = -\nabla^2 E = -\mu_R \varepsilon_R \frac{\partial^2 E}{\partial t^2}$$

- With the usual lossless plane-wave solution

$$E = E_0 e^{-j\beta x} e^{j\omega t} = E_0 e^{-j\omega x/u} e^{j\omega t} = E_0 e^{-j\omega x/v_p} e^{j\omega t}, \quad \beta = \omega \sqrt{\mu_R \varepsilon_R}$$

- Where $\beta = \omega/u$ is the wavenumber, ω is frequency in rad/s, and $u = v_p = (\varepsilon_R \mu_R)^{-1/2}$ is the phase velocity in m/s.

Metamaterial "Left-Handed" Transmission Lines

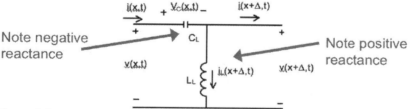

- Consider a lumped-element model of a "left-handed" line above
- The left-handed transmission line equations are

$$\frac{\partial^2 v(x,t)}{\partial x \partial t} = -\frac{1}{C_L} i(x,t) \quad \Rightarrow \quad \frac{\partial}{\partial t} \nabla \times E = -\frac{1}{\varepsilon_L} H$$

C is analogous to ε_L

$$\frac{\partial^2 i(x,t)}{\partial x \partial t} = -\frac{1}{L_L} v(x,t) \quad \Rightarrow \quad \frac{\partial}{\partial t} \left(\nabla \times H \right) = \frac{1}{\mu_L} E$$

L is analogous to μ_L

- The circuit model on the left corresponds to the three-dimensional Maxwell equations on the right
- Units are L_L in H·m and C_L in F·m

Propagation in "Left-Handed" Transmission Lines

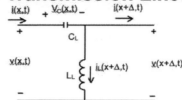

$$\frac{\partial}{\partial t}\nabla \times E = -\frac{1}{\varepsilon_L}H$$

$$\frac{\partial}{\partial t}\left(\nabla \times H\right) = \frac{1}{\mu_L}E$$

- Taking the curl and $\partial/\partial t$ of both sides

$$\frac{\partial}{\partial t}\nabla \times \left(\frac{\partial}{\partial t}\nabla \times E = -\frac{1}{\varepsilon_L}H\right) \Rightarrow \frac{\partial^2}{\partial t^2}\nabla \times \nabla \times E = -\frac{1}{\varepsilon_L}\frac{\partial}{\partial t}\nabla \times H$$

- Then, the wave equation follows as (where $\nabla \cdot E = 0$):

$$\frac{\partial^2}{\partial t^2}\nabla^2 E = \frac{1}{\mu_L\varepsilon_L}E$$

- With the usual plane-wave solution

$$E = E_0 e^{-j\beta x}e^{j\omega t} = E_0 e^{-j\omega x/u}e^{j\omega t}, \quad \beta = 1/\left(\omega\sqrt{\mu_L\varepsilon_L}\right)$$

- Where $\beta = \omega/u$ is the wavenumber, ω is frequency in rad/s, and $u = v_p$ is the phase velocity in m/s
- Next: solve for velocity u

"Left-Handed" and "Right-Handed" Solutions

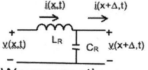

- Wave equations

$$\nabla^2 E = \mu_R\varepsilon_R\frac{\partial^2 E}{\partial t^2}$$

$$\frac{\partial^2}{\partial t^2}\nabla^2 E = \frac{1}{\mu_L\varepsilon_L}E$$

- Then, plug in general solution for both wave equations

$$E = E_0 e^{-j\beta x}e^{j\omega t} = E_0 e^{-j\omega x/u}e^{j\omega t}$$

- And get the phase velocities u_R and u_L and the solutions

$$u_R = \frac{1}{\sqrt{\mu_R\varepsilon_R}}$$

$$u_L = \omega^2\sqrt{\mu_L\varepsilon_L}$$

$$\Rightarrow E_0 e^{-j\omega x\sqrt{\mu_R\varepsilon_R}}e^{j\omega t}$$

$$\Rightarrow E_0 e^{-\frac{jx}{\omega\sqrt{\mu_L\varepsilon_L}}}e^{j\omega t}$$

Right-handed solution

Left-handed solution

"Left-Handed" and "Right-Handed" Group Velocities

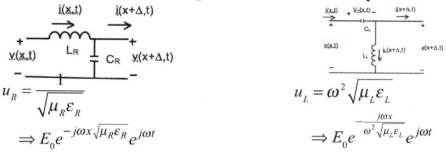

$$u_R = \frac{1}{\sqrt{\mu_R \varepsilon_R}}$$

$$\Rightarrow E_0 e^{-j\omega x \sqrt{\mu_R \varepsilon_R}} e^{j\omega t}$$

$$u_L = \omega^2 \sqrt{\mu_L \varepsilon_L}$$

$$\Rightarrow E_0 e^{-\frac{j\omega x}{\omega^2 \sqrt{\mu_L \varepsilon_L}}} e^{j\omega t}$$

- The group velocities u_{Rg} and u_{Lj} are $1/(dk/d\omega)$ where $\beta = \omega/u$

$$u_{Rg} = \frac{1}{d\beta/d\omega} = \frac{1}{d(\omega/u_R)/d\omega} =$$

$$= \frac{1}{d(\omega\sqrt{\mu_R \varepsilon_R})/d\omega} = \frac{1}{\sqrt{\mu_R \varepsilon_R}}$$

$$u_{Lg} = \frac{1}{d\beta/d\omega} = \frac{1}{d(\omega/u_L)/d\omega} =$$

$$= \frac{1}{d(\omega/\{\omega^2 \sqrt{\mu_L \varepsilon_L}\})/d\omega} = -\omega^2 \sqrt{\mu_L \varepsilon_L}$$

Right-handed solution   Left-handed solution

Note: group velocity gives the direction of energy flow

Comparison: "Left-Handed" and "Right-Handed" Velocities

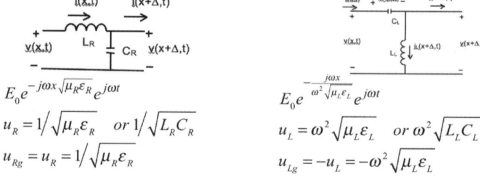

$$E_0 e^{-j\omega x \sqrt{\mu_R \varepsilon_R}} e^{j\omega t}$$

$$u_R = 1/\sqrt{\mu_R \varepsilon_R} \quad or \quad 1/\sqrt{L_R C_R}$$

$$u_{Rg} = u_R = 1/\sqrt{\mu_R \varepsilon_R}$$

$$E_0 e^{-\frac{j\omega x}{\omega^2 \sqrt{\mu_L \varepsilon_L}}} e^{j\omega t}$$

$$u_L = \omega^2 \sqrt{\mu_L \varepsilon_L} \quad or \quad \omega^2 \sqrt{L_L C_L}$$

$$u_{Lg} = -u_L = -\omega^2 \sqrt{\mu_L \varepsilon_L}$$

- Right-handed solution
- The velocities are constant
- The group and phase velocities are equal

- Left-handed solution
- The velocities change with frequency
- Group and phase velocities have opposite sign
- Called a "backward wave"

Note: group velocity gives the direction of energy flow

Analogy to Double-Negative Materials

Opposite sign series reactance

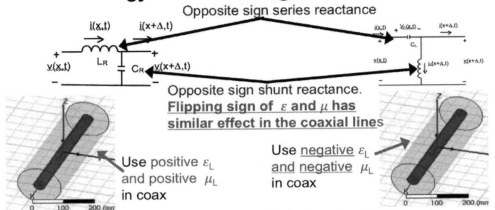

Opposite sign shunt reactance.
**Flipping sign of ε and μ has
similar effect in the coaxial lines**

Use positive ε_L
and positive μ_L
in coax

Use <u>negative</u> ε_L
and <u>negative</u> μ_L
in coax

- **Right-handed** LC line
- Analogous to a "**Normal**"
  transmission line
- Both permittivity ε_R and
  permeability μ_R are positive

- **Left-handed** LC line
- Analogous to a "**double-
  negative**" metamaterial
  (DNG)
- Both permittivity ε_L and
  permeability μ_L are negative

337

CRLH Structures

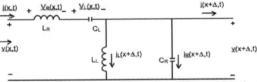

- A more common form for transmission lines is the composite
  right/left hand (CRLH) structure above
- This band-pass topology is left-handed at low frequencies
  and right-handed at high frequencies
- Analysis is along the same lines as before and would yield

$$\frac{\partial}{\partial t}\nabla \times E = -\mu_R \frac{\partial^2}{\partial t^2}H - \frac{1}{\varepsilon_L}H$$

$$\frac{\partial}{\partial t}\nabla \times H = \varepsilon_R \frac{\partial^2}{\partial t^2}E + \frac{1}{\mu_L}E$$

$$\Rightarrow u^2 = \frac{\omega^4 \mu_L \varepsilon_L}{\left(1-\omega^2\mu_R\varepsilon_L\right)\left(1-\varepsilon_R\mu_L\omega^2\right)}$$

See: Weldon, et al., "Left-Handed Extensions of Maxwell's Equations for
Metamaterials," IEEE SoutheastCon 2010 Proceedings, March 18-21, 2010.

Electromagnetic Metamaterials

Electromagnetic Metamaterials

- A variety of electromagnetic structures have been proposed for metamaterials that exhibit negative permeability or negative permittivity

- Typically metamaterials are formed by arrays of "unit cells"

- Key metamaterial unit cells include "split rings' and "I-beams'

- Most of these devices are narrowband because they depend on some underlying resonant mechanism

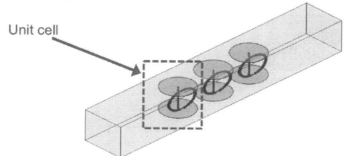

Unit cell

See: Weldon, et al., "A Wideband Microwave Double-Negative Metamaterial with Non-Foster Loading" IEEE SoutheastCon 2012 Proc., pp. March 15-18, 2012..

Single Split-Ring Resonator (SSRR)

- Split-ring resonators provide a narrowband metamaterial
- Ring current i_r is induced by an incident magnetic flux Φ_0
- Gap capacitance C_g and ring inductance L_R cause resonance

$$i_r = -\Phi_0 \frac{s^2 C_g}{1 + s^2 L_R C_g}; \quad \Phi_0 = \mu_0 H_0 A_R$$

Effective relative permeability

$$\mu_r = 1 - \frac{\mu_0 A_R^2}{l_x l_y l_z} \frac{s^2 C_g}{1 + s^2 L_R C_g}$$

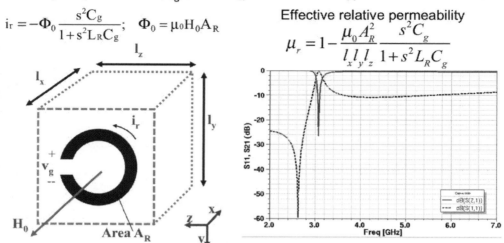

See: Weldon, et al., "A Wideband Microwave Double-Negative Metamaterial with Non-Foster Loading" IEEE SoutheastCon 2012 Proc., pp. March 15-18, 2012..

Non-Foster Wideband Single Split-Ring Resonator (SSRR)

- Unit cell is wideband with an inductance L_g added in the gap
- Ring current i_r is induced by an incident magnetic flux Φ_0
- For proper $L_g + L_R > 0$, metamaterial permeability $\mu_r < 0$

$$i_r = -\Phi_0 \frac{1}{L_R + L_g}; \quad \Phi_0 = \mu_0 H_0 A_R$$

Effective relative permeability

$$\mu_r = 1 - \frac{\mu_0 A_R^2}{l_x l_y l_z} \frac{1}{L_R + L_g}$$

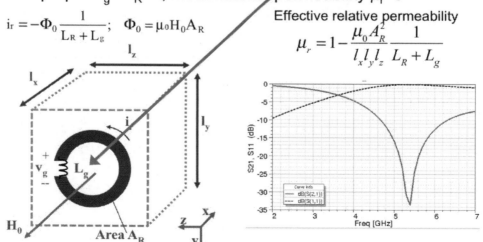

See: Weldon, et al., "A Wideband Microwave Double-Negative Metamaterial with Non-Foster Loading" IEEE SoutheastCon 2012 Proc., pp. March 15-18, 2012..

Electric Disk Resonator (EDR)

- EDR resonators provide a narrowband metamaterial
- Plate voltage v_d is induced by incident unit cell voltage v_0
- Post inductance L_p & fringe capacitance C_F cause resonance

$$v_d = v_0 \frac{s^2 L_p C_0}{1 + s^2 L_p C_F}$$

Effective relative permittivity

$$\varepsilon_r = 1 + \frac{A_d l_P}{l_x l_z (l_Y - l_P)} \left(\frac{1}{1 + s^2 L_p C_F} \right)$$

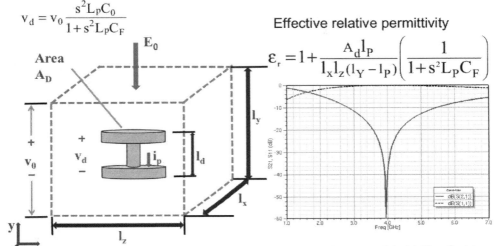

See: Weldon, et al., "A Wideband Microwave Double-Negative Metamaterial with Non-Foster Loading" IEEE SoutheastCon 2012 Proc., pp. March 15-18, 2012..

Non-Foster Wideband Electric Disk Resonator (EDR)

- Unit cell is wideband with capacitance C_p replacing the post
- Plate voltage v_d is induced by incident unit cell voltage v_0
- For proper $C_p + C_F > 0$, metamaterial permitivity $\epsilon_r < 0$

$$v_d = v_0 \frac{C_0}{C_p + C_F}$$

Effective relative permittivity

$$\varepsilon_r = 1 + \frac{C_0 l_P}{l_x l_z \varepsilon_0} \left(\frac{C_p}{C_p + C_F} \right)$$

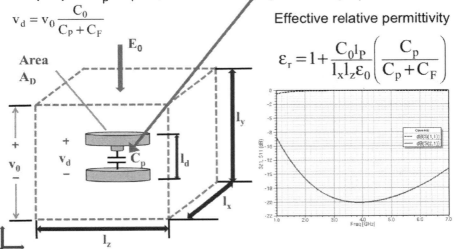

See: Weldon, et al., "A Wideband Microwave Double-Negative Metamaterial with Non-Foster Loading" IEEE SoutheastCon 2012 Proc., pp. March 15-18, 2012..

Electric Disk Resonator (EDR)

- Electromagnetic simulations can be used to observe the effective negative dielectric constant by observing the electric field reversal at frequency above resonance, as shown below
- The upper traces show the field when the unit cell is not present, for reference

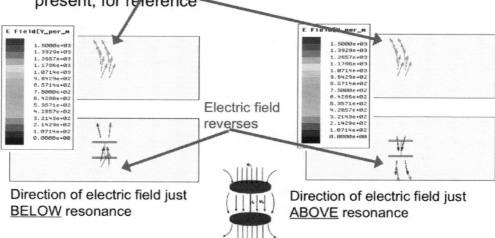

Electric field reverses

Direction of electric field just <u>BELOW</u> resonance

Direction of electric field just <u>ABOVE</u> resonance

Other unit Cells

- A wide variety of other unit cells exist
- Double-split ring is perhaps the most common type of split ring magnetic unit cell, and its planar form makes it easier to fabricate
- I-beam is a common electric unit cell, again with a planar form
- Double-split rings can also be rectangular, allowing tighter packing than circular rings
- More complex 3D shapes may useful in simulations for research
- Some complex 3D shapes may be simple enough to fabricate reasonably
- And many more...

Metamaterial Boundaries and Excitation Orientation

- Care must be taken to properly insert unit cells into structures with boundary conditions appropriate for a particular orientation of the unit cell

- As illustrated below, the excitation magnetic field is oriented to enter the split rings of the SSRRs and the electric field impinges on the disks of the EDRs

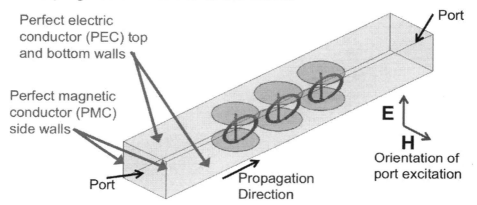

Perfect electric conductor (PEC) top and bottom walls

Perfect magnetic conductor (PMC) side walls

Port

Port

Propagation Direction

E

H

Orientation of port excitation

Setting Waveguide Excitations and De-Embedding

- For waveguide ports in some simulators, an "integration line" is used to orient the electric field excitation of a port as illustrated below

- This orientation should agree with the desired orientation of the metamaterial unit cells in the waveguide

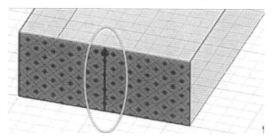

- Note: you must "de-embed" to the edge of your unit cells so that any phase/delay along empty waveguide portions does not cause error in estimating metamaterial properties

Extraction of Mu and Epsilon

- In our simulator, mu and epsilon of a metamaterial must be computed from s-parameters
- To extract mu and epsilon in the simulator, set up the following formulas (see lab website):
  - dd = length in meters of the physical region being characterized, such as the diameter of a split ring
  - v1 = S(1,1)+S(2,1)
  - v2 = S(2,1)-S(1,1)
  - k0 = 6.28*Freq/3e8
  - mur = 2/(cmplx(0,1)*k0*dd)*(1-v2)/(1+v2)
  - epsr = mur+2*cmplx(0,1)/(k0*dd)
  - epsr2 = 2/(cmplx(0,1)*k0*dd)*(1-v1)/(1+v1)
- For details, see: R.W. Ziolkowski, 'Design, fabrication, and testing of double negative metamaterials," IEEE Tran. on Antennas and Prop., 51:7, July 2003.

11 APPENDIX

.

Useful Vector Identities

$$\nabla \cdot (\nabla \times \mathbf{A}) = 0$$

$$\nabla \times (\nabla V) = 0 \quad \text{for any scalar potential V}$$

$$\nabla \cdot (\nabla V) = \nabla^2 V \quad \text{for any scalar potential V}$$

$$\nabla \cdot (\nabla \mathbf{E}) = \nabla^2 \mathbf{E} = \nabla(\nabla \cdot \mathbf{E}) - \nabla \times \nabla \times \mathbf{E} \quad \text{for any vector field } \mathbf{E}$$

$$\mathbf{A} \times (\mathbf{B} \times \mathbf{C}) = \mathbf{B}(\mathbf{A} \cdot \mathbf{C}) - \mathbf{C}(\mathbf{A} \cdot \mathbf{B})$$

$$\mathbf{A} \cdot \mathbf{B} = \mathbf{B} \cdot \mathbf{A} \qquad \mathbf{A} \cdot (\mathbf{B} + \mathbf{C}) = \mathbf{A} \cdot \mathbf{B} + \mathbf{A} \cdot \mathbf{C} \qquad \mathbf{A} \cdot \mathbf{A} = |\mathbf{A}|^2$$

$$\mathbf{A} \times \mathbf{B} = -\mathbf{B} \times \mathbf{A} \qquad \mathbf{A} \times (\mathbf{B} + \mathbf{C}) = \mathbf{A} \times \mathbf{B} + \mathbf{A} \times \mathbf{C} \qquad \mathbf{A} \times \mathbf{A} = 0$$

$$\mathbf{A} \cdot \mathbf{B} = \mathbf{A}^T \mathbf{B} = |\mathbf{A}||\mathbf{B}| \cos(\theta) = A_x B_x + A_y B_y + A_z B_z$$

Trigonometry Identities

$$cos(A)\, cos(B) = 0.5\, cos(A - B) + 0.5\, cos(A + B)$$

$$sin(A)\, cos(B) = 0.5\, sin(A - B) + 0.5\, sin(A + B)$$

$$sin(A)\, sin(B) = 0.5\, cos(A - B) - 0.5\, cos(A + B)$$

$$\cos^2(A) = 0.5 + 0.5 \cos(2A)$$

$$\sin^2(A) = 0.5 - 0.5 \cos(2A)$$

$$a\cos(x) + b\sin(x) = \sqrt{a^2 + b^2}\, \cos(x + \arctan(-b/a))$$

$$e^{j\theta} = \cos(\theta) + j\, \sin(\theta)$$

$$e^{j\theta} + e^{-j\theta} = 2\cos(\theta)$$

$$e^{j\theta} - e^{-j\theta} = 2j\sin(\theta)$$

Quadratic formula:

$$ax^2 + bx + c = 0 \quad \Rightarrow \quad x = \frac{-b \pm \sqrt{b^2 - 4ac}}{2a}$$

Laplace Transforms and Z-Transforms

Laplace Trans. $X(s)$	Contin. Time $x(t)$	Sampled Function $x[n]=x(nT_s)$	z-transform. $X(z)$
1	$\delta(t)$	–	–
$\dfrac{1}{s}; \operatorname{Re}\{s\}>0$	$u(t)$	$u[n]$	$\dfrac{z}{z-1}; \ |z|>1$
$\dfrac{1}{s^2}; \operatorname{Re}\{s\}>0$	$tu(t)$	$nT_s u[n]$	$\dfrac{zT_s}{(z-1)^2}; \ |z|>1$
$\dfrac{1}{(s+a)}; \operatorname{Re}\{s\}>-a$	$e^{-at}u(t)$	$e^{-naT_s}u[n]$	$\dfrac{z}{z-e^{-aT_s}}; \ |z|>|e^{-aT_s}|$
$\dfrac{1}{(s+a)^2}; \operatorname{Re}\{s\}>-a$	$te^{-at}u(t)$	$nT_s e^{-naT_s}u[n]$	$\dfrac{ze^{-aT_s}T_s}{\left(z-e^{-aT_s}\right)^2}; \ |z|>|e^{-aT_s}|$
$\dfrac{a}{s(s+a)}; \operatorname{Re}\{s\}>0$	$(1-e^{-at})u(t)$	$\left(1-e^{-naT_s}\right)u[n]$	$\dfrac{z\left(1-e^{-aT_s}\right)}{(z-1)\left(z-e^{-aT_s}\right)}; \ |z|>1$
$\dfrac{s}{s^2+\Omega_0^{\ 2}}; \operatorname{Re}\{s\}>0$	$\cos(\Omega_0 t)u(t)$	$\cos(n\Omega_0 T_s)u[n]$	$\dfrac{z^2-z\cos(\Omega_0 T_s)}{z^2-2z\cos(\Omega_0 T_s)+1}; \ |z|>1$
$\dfrac{\Omega_0}{s^2+\Omega_0^{\ 2}}; \operatorname{Re}\{s\}>0$	$\sin(\Omega_0 t)u(t)$	$\sin(n\Omega_0 T_s)u[n]$	$\dfrac{z\sin(\Omega_0 T_s)}{z^2-2z\cos(\Omega_0 T_s)+1}; \ |z|>1$

Z-transform Pairs

Discrete-time Function	z-transform
$\delta[n]$	1
$u[n]$	$\dfrac{z}{z-1};\ \ \|z\|>1$
$nu[n]$	$\dfrac{z}{(z-1)^2};\ \ \|z\|>1$
$a^n u[n]$	$\dfrac{z}{z-a};\ \ \|z\|>\|a\|$
$-a^n u[-n-1]$	$\dfrac{z}{z-a};\ \ \|z\|<\|a\|$
$na^n u[n]$	$\dfrac{az}{(z-a)^2};\ \ \|z\|>\|a\|$
$\cos(\omega_0 n)u[n]$	$\dfrac{z^2 - z\cos(\omega_0)}{z^2 - 2z\cos(\omega_0)+1};\ \ \|z\|>1$
$\sin(\omega_0 n)u[n]$	$\dfrac{z\sin(\omega_0)}{z^2 - 2z\cos(\omega_0)+1};\ \ \|z\|>1$

Modified Z-Transforms

Mod. z-transform, X(z,m)	z-transform	Time Func	Laplace Transform
	1	$\delta(t)$	1
$\dfrac{1}{z-1}$	$\dfrac{z}{z-1};\ \ \|z\|>0$	$u(t)$	$1/s$
$\dfrac{mT_0}{z-1}+\dfrac{T_0}{(z-1)^2}$	$\dfrac{zT_0}{(z-1)^2};\ \ \|z\|>1$	$tu(t)$	$1/s^2$
$\dfrac{e^{-amT_0}}{z-e^{-aT_0}}$	$\dfrac{z}{z-e^{-aT_0}};\ \ \|z\|>e^{-aT_0}$	$e^{-at}u(t)$	$\dfrac{1}{s+a}$
$\dfrac{T_0 e^{-amT_0}\left[e^{-aT_0}+m(z-e^{-aT_0})\right]}{(z-e^{-aT_0})^2}$	$\dfrac{T_0 z e^{-aT_0}}{(z-e^{-aT_0})^2};\ \ \|z\|>e^{-aT_0}$	$te^{-at}u(t)$	$\dfrac{1}{(s+a)^2}$

Fourier Transform in "f"

Fourier Transform Pairs $\quad X(f) = \int_{-\infty}^{\infty} x(t)e^{-j2\pi ft}dt \qquad x(t) = \int_{-\infty}^{\infty} X(f)e^{j2\pi ft}df$

$\delta(t) \leftrightarrow 1$	$\cos(2\pi f_0 t) \leftrightarrow 0.5(\delta(f + f_0) + \delta(f - f_0))$
$1 \leftrightarrow \delta(f)$	$\sin(2\pi f_0 t) \leftrightarrow 0.5j(\delta(f + f_0) - \delta(f - f_0))$
$u(t) \leftrightarrow \dfrac{1}{2}\delta(f) + \dfrac{1}{j2\pi f}$	$sgn(t) \leftrightarrow \dfrac{1}{j\pi f}$
$\Pi(t/\tau) \leftrightarrow \tau \cdot sinc(\pi f\tau)$	$2B\, sinc(2\pi Bt) \leftrightarrow \Pi(f/(2B))$
$\Delta\left(\dfrac{t}{\tau}\right) \leftrightarrow \dfrac{\tau}{2} \cdot sinc^2(\pi f\tau/2)$	$B\, sinc^2(\pi Bt) \leftrightarrow \Delta(f/(2B))$
$e^{j2\pi f_0 t} \leftrightarrow \delta(f - f_0)$	$e^{-at}u(t) \leftrightarrow \dfrac{1}{a + j2\pi f}$
$e^{-t^2/(2\sigma^2)} \leftrightarrow \sigma\sqrt{2\pi}\, e^{-2(\sigma\pi f)^2}$	$e^{-a|t|} \leftrightarrow \dfrac{2a}{a^2 + (2\pi f)^2}$

Fourier Transform Properties

$g(t)e^{j2\pi f_0 t} \leftrightarrow G(f - f_0)$	$g(t - t_0) \leftrightarrow G(f)e^{-j2\pi t_0 f}$
$g(at) \leftrightarrow \dfrac{1}{|a|}G\left(\dfrac{f}{a}\right)$	$G(t) \leftrightarrow g(-f)$
$g(t) * h(t) \leftrightarrow G(f)H(f)$	$g(t)h(t) \leftrightarrow G(f) * H(f)$
$\dfrac{dg(t)}{dt} \leftrightarrow j2\pi f\, G(f)$	$\displaystyle\int_{-\infty}^{t} g(\alpha)d\alpha \leftrightarrow \dfrac{G(f)}{j2\pi f} + G(0)\delta(f)/2$
	$\displaystyle\int_{-\infty}^{\infty} |g(t)|^2 dt = \int_{-\infty}^{\infty} |G(f)|^2 df$

Fourier Transform in "Ω"

Fourier Transform Pairs $\quad X(\Omega) = \int_{-\infty}^{\infty} x(t)e^{-j\Omega t}dt \qquad x(t) = \frac{1}{2\pi}\int_{-\infty}^{\infty} X(\Omega)e^{j\Omega t}d\Omega$

$\delta(t) \leftrightarrow 1$	$\cos(\Omega_0 t) \leftrightarrow \pi(\delta(\Omega + \Omega_0) + \delta(\Omega - \Omega_0))$
$1 \leftrightarrow 2\pi\delta(\Omega)$	$\sin(\Omega_0 t) \leftrightarrow j\pi(\delta(\Omega + \Omega_0) - \delta(\Omega - \Omega_0))$
$u(t) \leftrightarrow \pi\delta(\Omega) + \dfrac{1}{j\Omega}$	$sgn(t) \leftrightarrow \dfrac{2}{j\Omega}$
$\Pi(t/\tau) \leftrightarrow \tau \cdot sinc(\Omega\tau/2)$	$W\, sinc(Wt) \leftrightarrow \pi\, \Pi(\Omega/(2W))$
$\Delta\left(\dfrac{t}{\tau}\right) \leftrightarrow \dfrac{\tau}{2} \cdot sinc^2(\Omega\tau/4)$	$W\, sinc^2(Wt/2) \leftrightarrow 2\pi\Delta(\Omega/(2W))$
$e^{j\Omega_0 t} \leftrightarrow 2\pi\delta(\Omega - \Omega_0)$	$e^{-at}u(t) \leftrightarrow \dfrac{1}{a + j\Omega}$
$e^{-t^2/(2\sigma^2)} \leftrightarrow \sigma\sqrt{2\pi}\, e^{-\sigma^2\Omega^2/2}$	$e^{-a|t|} \leftrightarrow \dfrac{2a}{a^2 + \Omega^2}$

Fourier Transform Properties

$g(t)e^{j\Omega_0 t} \leftrightarrow G(\Omega - \Omega_0)$	$g(t - t_0) \leftrightarrow G(\Omega)e^{-jt_0\Omega}$
$g(at) \leftrightarrow \dfrac{1}{|a|}G\left(\dfrac{\Omega}{a}\right)$	$G(t) \leftrightarrow 2\pi g(-\Omega)$
$g(t) * h(t) \leftrightarrow G(\Omega)H(\Omega)$	$g(t)h(t) \leftrightarrow \dfrac{1}{2\pi}G(\Omega) * H(\Omega)$
$\dfrac{dg(t)}{dt} \leftrightarrow j\Omega\, G(\Omega)$	$\displaystyle\int_{-\infty}^{t} g(\alpha)d\alpha \leftrightarrow \dfrac{G(\Omega)}{j\Omega} + \pi G(0)\delta(\Omega)$
	$\displaystyle\int_{-\infty}^{\infty}|g(t)|^2 dt = \dfrac{1}{2\pi}\int_{-\infty}^{\infty}|G(\Omega)|^2 d\Omega$

Maxwell's Equations (Shorthand)

Point Form (Differential form)   Integral Form

$$\nabla \times \mathbf{E} = -\frac{\partial \mathbf{B}}{\partial t}$$
Faraday's Law
$$\oint_L \mathbf{E} \cdot d\mathbf{L} = -\int_S \frac{\partial \mathbf{B}}{\partial t} \cdot d\mathbf{S}$$

assuming $\mathbf{J}_m = 0$

$$\nabla \times \mathbf{H} = \mathbf{J}_e + \frac{\partial \mathbf{D}}{\partial t}$$
Ampere
Circuital Law
$$\oint_L \mathbf{H} \cdot d\mathbf{L} = \int_S \left(\mathbf{J}_e + \frac{\partial \mathbf{D}}{\partial t} \right) \cdot d\mathbf{S}$$

$$\nabla \cdot \mathbf{D} = \rho_e$$
Gauss' Law
$$Q = \int_V \rho_e \, dv = \oint_S \mathbf{D} \cdot d\mathbf{S}$$

$$\nabla \cdot \mathbf{B} = 0$$
Gauss' Magnetism Law
$$0 = \oint_S \mathbf{B} \cdot d\mathbf{S}$$

assuming $\rho_m(\overline{x}, t) = 0$

$\mathbf{D} = \varepsilon \mathbf{E}$ if $\varepsilon(t) = \delta(t)\varepsilon$ $\mathbf{J}_e = \sigma \mathbf{E}$

$\mathbf{B} = \mu \mathbf{H}$ if $\mu(t) = \delta(t)\mu$ and $\overline{x} = [x \ y \ z]^T$

otherwise

$\mathbf{D} = \varepsilon(t) * \mathbf{E}$

$\mathbf{B} = \mu(t) * \mathbf{H}$

Phasor Form of Maxwell's Equations

- The point and integral phasor forms

Point Form (Differential form)   Integral Form

$$\nabla \times \mathbf{E} = -j\omega \mathbf{B}$$
Faraday's Law
$$\oint_L \mathbf{E} \cdot d\mathbf{L} = -j\omega \int_S \mathbf{B} \cdot d\mathbf{S}$$

assuming $\mathbf{J}_m = 0$

$$\nabla \times \mathbf{H} = \mathbf{J}_e + j\omega \mathbf{D}$$
Ampere
Circuital Law
$$\oint_L \mathbf{H} \cdot d\mathbf{L} = \int_S \left(\mathbf{J}_e + j\omega \mathbf{D} \right) \cdot d\mathbf{S}$$

$$\nabla \cdot \mathbf{D} = \rho_e$$
Gauss' Law
$$Q = \int_V \rho_e \, dv = \oint_S \mathbf{D} \cdot d\mathbf{S}$$

$$\nabla \cdot \mathbf{B} = 0$$
Gauss' Magnetism Law
$$0 = \oint_S \mathbf{B} \cdot d\mathbf{S}$$

assuming $\rho_m(\overline{x}, t) = 0$

$\mathbf{D} = \varepsilon \mathbf{E}$ $\mathbf{J}_e = \sigma \mathbf{E}$

$\mathbf{B} = \mu \mathbf{H}$

Test of Symbol Fonts

- Arial symbols:
- ©®Ωℑℜℱℤ⊕✕×⊗∩∪⊕⊗≈∧≥≦≤Δ∇∀∞
- →→⟹⇔×÷±∓≈≠∘‥●∗·√∑∫∠|∮∮∮
- αβγδεζηθικλμνξοπρςστυφχψω ∂εϑϰϕϱϖ**A**
- ΑΒΓΔΕΖΗΘΙΚΛΜΝΞΟΠΡΘΣΤΥΦΧΨΩ ∇

Made in the USA
Columbia, SC
10 January 2020